The Road To 59

THE ROAD TO 59

By T. Ellis

The Road To 59

Dedicated

to

all the players, teams, and fans who have not yet experienced a World Championship

"Luck is the reward for having faith, & believing in the impossible, and doing everything in your power to make that happen"

The Road To 59

T.Ellis
The Road to 59

© 2021,T.Ellis
Self-published

ISBN: 9781737108306

info@theroadto59.com

© 2021 All rights reserved.

No part of this publication may be reproduced, stored in a retrieval system, stored in a database and/or published in any form or by any means, electronic, mechanical, photo copying, recording or otherwise, without the prior written permission of the publisher.

TABLE OF CONTENTS

Introduction	2
The Chance Encounter	6
His Beginnings	12
The List	22
The Story	30
Good Luck	49
Bad Luck	61
Quotes	69
Sports Trivia	75
The Follow-up	86
List of Teams	92
Bucket List	139
World Wide Contest	151
Bibliography	153

INTRODUCTION

Introduction

Does God exist? If so, is there any proof? Can you see him? Who said God is a he? What if he is actually a she? Does she like sports? If so, is she a football fan? The world sport is soccer (football), so she must love soccer. You can't see God, you can't talk to God, you can't touch God, but billions of people believe in God. You can't see atoms, you can't touch them, but we have proof that they exist. You can see players pray and give God their praise when they win and say nothing when they lose. You read about athletes following certain routines that seem very idiotic to the normal person but have great meaning to the athlete and how it will affect his or her game. Some call it divine intervention when it works, and others call it just plain luck. Whichever one it is, it's still very superstitious, and If it works, does it matter what we call it?

Remember that good luck does not need an explanation; just be thankful that it is on your side. Believing in good luck has proven to make athletes play better, and in doing so makes their chances of winning the game more likely. To the best of my ability, I will try to tell a story of unbelievable luck for just one person to have, and how it has translated into the wide world of sports.

What you will read about here is the fifty-eight world championships that one man has brought to players, coaches, and fans from around the world for over thirty-two years—and who has done it with no media attention, no rewards, and no contracts, expecting nothing in return. If you are a fan, then this is the guy you want in your corner so that your favorite team can win a championship. It is like that movie, Good Luck Chuck, with Jessica Alba and Dane Cook. Any woman who dates the character played by Dane Cook meets her true love with the next person she dates. It's a great comedy and very relevant to the story I am about to tell you. This guy goes on an assignment in a city, a state, or a country, and what follows is a championship, bringing luck to that city's team like Chuck brings luck to his dates in the movie.

I will leave you with an analogy to give you some insight into how good luck works. Imagine you are a sniper in the United States military. You are the best of the best. You have trained for years and have outperformed all other soldiers in your field of expertise. You hit your targets ninety-nine percent of the time. With your talent and training, you are one of the best in the world. One day you are out on a mission somewhere in the world and you find yourself up against another sniper who's just as good as you are, maybe even better. Your heart is beating fast; your adrenaline is up a notch or two. You are looking for him, but he finds you first and has you in his sights; his crosshairs are in the center of your head. He feels comfortable and takes the shot, but he misses you.

What are your thoughts as that bullet flies by your head? I can tell you, and any other combat soldier can tell you, what your thoughts would be: "I am damn lucky." Because somewhere deep inside of you, your gut, your instinct, your conscience, you know he should have never missed because you would have never missed. Ask any combat soldier whether he or she would rather be lucky or good. Lucky gets you back home to your spouse, to your children. It gets you back alive. You can have all the talent and all the training in the world, but luck will hopefully be there when you need it the most. Just like in sports, a team can be the most talented and have a great coaching staff, but they need a little luck sometimes—or maybe more often than sometimes. Just ask the 2008 New England Patriots, who were favored to win that Super Bowl and complete an undefeated season. But wait! Was it luck or bad luck that intervened and snatched that perfect season away from the heavily favored Patriots?

We all remember the famous "helmet catch." If you are a Giants fan, then you believe in good luck, and if you are a Patriots fan, then you see it as bad luck. Who decides which teams have good luck or bad luck? I will tell you explicitly in this book. I will prove to you, through science and through faith, that good luck and bad luck coexist uniformly and do so on the same playing field at the same time. I will show you the way to help your team win that world championship that has so eluded you and your team for so long. In the end, it is your faith and your team's talent and good luck that will prevail and will prove to you and the world that good luck exists. It exists in one man, and one man only. This is his story.

If you take this journey with me, I will assure you that your team will win a world championship. I promise you that there will be no disappointment at the end of the season. The story I am about to tell you is a true story, it's a heartfelt story. It's a story of immense luck and how one man's journey has led to fifty-eight national and world championships.

"Remember luck is the reward for your faith and belief"

—T. Ellis

THE CHANCE ENCOUNTER

Chapter I
The Chance Encounter

"Good luck needs no explanation."
—Shirley Temple, child actress

"The amount of good luck coming your way depends on your willingness" to act.
—Barbara Sher, bestselling author

This story begins on my birthday back on December 21, 2016. It was freezing and snowing as it always is on Minnesota winter days. It was perfect for the first day of winter and even better for my birthday celebration. My friends and I had gathered at a local establishment that we frequented often throughout the year, especially during the summer, and on that night, no snow or freezing temperatures were going to keep me from celebrating. That night was special because it was not just my birthday. It would become a much greater night of importance than I realized at the time because I met him on that night.

I am a social person who likes to talk to anybody who has a good and interesting conversation going on, and that is when I stumbled upon the stranger who would change my life and my birthday celebration. I was up at the bar ordering a round of drinks for my group when I eavesdropped on a conversation that caught my attention because it had to do with sports. I am a big sports fan because I was an athlete since I was in second grade. I played football, wrestled, boxed, and competed in track and field as my go-to sports but also developed an interest in following multiple spectator sports.

The gentleman next to me was asking people around him if they could answer a sports trivia question. Well, of course I was interested in trying to figure it out. I love trying to solve trivia questions. As I was at the bar ordering some drinks and flirting with the bartender a little, I tried to answer the question, but I could not get the name of the last team to complete the trivia answer. Maybe because I was a little more intoxicated than I realized... No one at the bar could solve the challenge. Well, not without cheating and glancing at their smartphones. The trivia question seemed simple enough until you thought about it. For the life of me, I could not come up with the name of the last team.

"Can you tell me the question again?" I urged this stranger, and he obliged me and the rest of the pack who had gathered around him. He had some of the toughest sports trivia questions I had ever heard, not like the statistics that sports analysts spew out, but the kind that are entertaining even if it's hopeless sports knowledge. He restated the question to all of us: "Among all the major professional teams in the NFL, NBA, MLB, and NHL, what teams do not end with the letter (s)?" He used a college team, the University of Alabama Crimson Tide, as an example that does not end in the letter (s). "There are eight teams that match this criterion in professional sports, so good luck."

The night continued on, and I realized this stranger was brilliant about sports—I mean, all sports. Still, no one could solve the trivia question, so we called for the answers. Here they are:

1. Boston Red Sox

2. Chicago White Sox

3. Colorado Avalanche

4. Miami Heat

5. Minnesota Wild

6. Oklahoma Thunder

7. Orlando Magic

8. Tampa Bay Lightning

9. Utah Jazz

The one team I could not remember was the Tampa Bay Lightning. OMG, I was so close. He was a sports guy, but no one knew who he was, just some guy in a bar drinking and socializing with the other patrons. I was interested in talking with him more because I was a big sports fan, and I liked all of his sports trivia questions. I left my friends for a time so I could chat with this stranger. After introducing myself to him, he did the same in return. I learned he was a combat veteran and a traveling nurse. We had a lot in common, and I learned as well about his love for sports. This very weird story of him being a real, live good luck charm intrigued me. Thinking to myself that he might be a little drunk, I was still interested in knowing more.

As the night continued, I gained a lot of insight about this gentleman, and I appreciated his stories and his charm. My companions reached over to see if I was ready to leave and to remind me we needed to get going to the next spot to continue the birthday celebration. Not wanting to leave because I was so enthralled by this stranger's story, I finally did so after some convincing by my friends. I invited this stranger to join us, but he replied, "no thank you," so I offered to exchange contact information, and he agreed to that. As the night flew by, I tried his sports trivia quiz on complete strangers at the various bars we stumbled into, and they were colossal hits with most of the patrons we came across. It showed that useless knowledge is popular, especially when you are drunk. As the night kept going, I became more intoxicated and louder, and sooner or later ended up safely at home in bed thanks to my good friends.

Waking up at about 11 a.m., I really had to pee and required water, lots of water. Reliving the night before, all the shots and flirting, I thought of the stranger I met with the sports trivia knowledge and his story about being a real, live good luck charm. It engrossed me. Call it simple curiosity or professional curiosity (because he was a fellow military man), but it picked at me all day. To quell this curiosity, I called him to see if he would meet me for lunch. But first, I asked about his story on the phone. I told him I wanted to write an article about his good luck with the sports team's winning championships. He agreed to meet me on one condition—that I would leave his name out of the article. He said I could use the facts surrounding the good luck but no mention of his name. I agreed, and we met later that day for a late lunch.

I met him at a local sports bar in the 7 Corners neighborhood, close to the University of Minnesota. It was calm during the holidays, so it would be a suitable spot for me to conduct the interview with him. He showed up a little later as he was coming from an appointment and overestimated how long it would take to make it there. When he finally arrived, he walked right over to my table, squatted down, and asked me what I wanted to know. I replied, "Everything. Why do you consider yourself a good luck charm?"

"MIRACLES ARE MANY THINGS TO MANY DIFFERENT PEOPLE"

Chapter II
His Beginnings

"Believe, don't believe. The choice is yours."
—Unknown

Here, to the best of my recollection, is what he said to me that day:

"Starting with when my mother was pregnant with me, let me start by telling you why I think good luck has accompanied me my entire life. Well, at least so far, it has. This good luck started before I was even born."

"My mother gave birth to me at six-and-a-half months instead of the usual nine months because of an injury caused by her brother. That is correct, I said her brother. My uncle pushed my mother down the stairs, which induced her to give birth to me prematurely. He was not a hateful man, but an angry man. He was a Green Beret in Vietnam. That can be rough on a man. When I was younger, visiting his home with my mother, I recall looking at this photograph on the wall next to all the medals he and his team earned in Vietnam. He was holding up the head of a North Vietnamese. It was just the head, not the body. I still remember not being shocked or unsettled about the photograph. I am uncertain why. Well, I am not giving my uncle an excuse for what he did but trying to understand him and the condition he was in."

"He had almost killed me and my mother. When he shoved my mother down those stairs, it created a chain reaction for the three of us—me, my mother, and him. That action would transform our lives forever. For me, it was just the beginning of the good fortune and extremely good luck that I would have for the rest of my existence. For my mother, it would be a difficult and hard life with many mistakes along the way. I give my mother tremendous credit for providing and caring for four boys by herself. She did the best she could with not a lot of support from family members. For my uncle, it would be a life of regret, guilt, and drinking."

"My life started out as a baby so small that I could fit in a woman's shoebox. I was in an incubator for over three months. The reason my uncle shoved my mother down the stairs was because he thought she was pregnant by a black man. I suppose that in 1968, that was heavily frowned on. My father was a white Irishman, so maybe that might be why I am so lucky haha. So anyways my mother never forgave her brother or absolved him for that event. It led to a split within our family that lingers to this very day. My grandmother called me the 'miracle baby' for having survived those first three months, not to mention the first year. What's ironic about the situation is that my uncle ended up adopting an African-American child."

"My birth was miracle number one. Miracle number two was indeed scarier. I was four years old, and like any youthful child, I enjoyed playing with my toys. I had this Fisher-Price construction set: a plastic hammer, screwdriver, helmet, and belt. I was playing with the hammer, pretending to build something. I think I was struggling to make a wall, unclear who was going to pay for it, LOL. I doubted my mother was going to pay for it. Instead of building something, however, I ended up falling against and shattering the window along with my left ulnar and radial nerve. The glass lacerated my arm from wrist to mid-forearm. My mother clutched a towel to stop the bleeding. We lived out in the country, so the ride to the emergency room was long. I recall this next part in great detail. When my mother was driving me to the hospital, I was crying like a banshee, and she ordered me to look at the cows in the fields as we sped by them. She kept telling me to look at the cows, look at the cows, and when I did, I stopped crying. I will never forget that. My mother was so calm and strong. In the present, when I take my dog for a car ride and I instruct him to go look at the cows, he charges to the window to look for one."

"I was left-handed, so after the accident, I had to learn to write with my right hand. Talk about a rough start to life! I can hardly imagine what my mother had to go through with me. I feel I am to blame for my mother's burdens. Being a mother is not a smooth role to have. The surgeon pointed out to my mother that it was a miracle that I survived that accident. I had lost a lot of blood. That a child that age and that size survived such an injury, he said, was a miracle. That was miracle number two."

"Miracle number three was the scariest for me. Being strangled to death, suddenly being revived, is something I cannot fully describe. There was a neighbor boy I was friends with. He had a terrible temper that exploded onto me one day. I remember what I felt to be a white light, the type that people describe feeling at the point of near-death. It was like being brushed over by white paint just as your soul is being sapped from your body. It was the scariest thing I have ever gone through in my life so far. I passed out, or as my neighbor's dad called it, I 'came back from the dead.' He was the one who performed CPR and claimed I did not have a pulse when he checked for one."

"Looking back on it, I understand that choking to death is one of the scariest ways to die. You realize you can't do anything to stop it, and you feel the life leaving your body. Your soul is being snuffed out as you stare at the perpetrator carrying out the crime against you and wonder why he is doing this to you. I did not despise him for it, but felt sorry for him. I felt his anger right at that moment, and suddenly I felt sorry for him because he did not know what he was doing. It was sheer rage, nothing else. I will never forget that feeling. I saw him later in life, and we spoke, and I forgave him."

"Miracle number four was a pileup that led to a hole in my right patella. I also suffered a concussion, dislocated my left shoulder, and flung up into the midst of traffic. The witnesses said I was very lucky I did not get run over after being thrown over twenty feet in the air. The witnesses stated that I looked like a pinball bouncing against the bumpers. One witness described it as my body being lifted in the air, ricocheting off of one car and into another, and suddenly laying limp on the ground. The paramedic said that I struggled to get up and walk away but stumbled back down, causing a concussion."

"I had to have pins put in my patella until the bone grew back. My physical therapist was a very exquisite young woman. I was extra motivated to become stronger, to show her I could and also to keep going there just so I could spend more time with her. I wrestled for seven years, ran track in high school, and joined the United States Army as an infantryman. The extra work paid off!"

"The next miracle was one that I have strong personal feelings about. I will just say that it included a friendly fire incident involving the United States Army and the mainstream media. I lost one of my best friends. I survived, and he did not. I suffer from survivor's guilt, and I have frequent nightmares. I miss my friend and think about him often. Miracle five was a genuine tragedy. It continues to haunt me."

"Miracle number six was after my time in the war. Everyone was celebrating and partying. My friends and I were down on the beach all night. Tybee Island, Georgia, was full of people from one end of the beach to the other end. There were young women everywhere, and they were more than willing to celebrate with us. My buddy and I had partied with these girls from Georgia all night. I spent the night with this pretty redhead on the beach. She was a delicious distraction from what I was feeling about the war. At that point, I was just happy to be alive and to be back in the States."

"After staying up all night and entertaining these girls, my buddy and I had to make it back to the base. We said our goodbyes, exchanged phone numbers, and started our trip back. As we were leaving, another soldier asked for a ride back to his base. Of course, it would be no trouble; soldiers always helped one another in such circumstances. He was stationed at a local Air Force base and it was on our way to where we had to go. Before we could drive through the AFB gates, we had to put our seatbelts on. Air Force security personnel are sticklers for the rules, but because they are, it saved our lives. I was driving a jeep, and when the tire blew as we were driving back to our base, the jeep flipped upside down. The only thing that kept us from having our heads decapitated was the seatbelts. Lucky or a miracle? What do you think?"

"Miracle number 7 is maybe more lucky than a miracle. You can decide if it merits the label miracle or just me being very lucky again. I had just gotten back from a military training school and was in great shape. My friends and I played football on Saturdays afternoons. You know how a group of friends always seem to have one asshole friend, the one they tolerate but would never allow to date their sister? You know the type I'm talking about."

"I have known this person since high school. He thought he was the baddest motherfucker on the field, but that was the furthest from the truth. I was friends with him starting in high school, but as time went by, I saw the truth about him, and I was trying to distance myself from his toxic lifestyle. This terrible person was a bully; he manipulated people, cheated on his wife. He was so horrible to his wife that I believe he caused her mental breakdown."

"When I came back from the military, like I said, I was in great shape. I had scored a 362 on my physical training test. A 300 is the best you can get, but they have an extended chart for soldiers that go over the maximum. So at the least, I was feeling great when I showed up for Saturday football. It was great to see everyone again and catch up with what we had been up to in our lives."

"We waited for everyone to show before we started the game. As soon as we had enough people, we began our ritual game of Saturday afternoon football. My asshole friend I was telling you about showed up and, like always, he had to go the extra step and not abide by the rules set up by all of us. I went up for a catch and he took my feet out from under me, and flipped me upside down. I hit the ground hard, and immediately my other friends came over to take this jerk down and give him a taste of his own medicine. I got up right away to go after him, and my friend stopped me in mid-stride. He whispered to me, 'It's not worth it.' I did not feel that way at all, but he held me until I calmed down. The rules were no-tackle, two-hand-touch only to keep injuries at a minimum."

"The game ended in my team winning 42-14. We all went out for drinks and food after the game to relax and calm the waters of any bad feelings about the game. The local bar we used to frequent was called was... OMG, I cannot remember what the name of the bar was. We used to go there every Saturday after the game. Well, it doesn't matter; it was the group of guys I played with that's important, and the time we spent afterwards."

"I know this next part has nothing to do with my good luck, but there was this guy we used to play with named Reidy. He was older than us by ten to twenty years. I'm not sure about his exact age, but he played with us all the time. He was the type of guy we all looked up to, the guy who had gravitas. Everybody loved him. He was well liked, and a lot of us wanted to be like him when we got older. I always wonder what happened to him. Just telling this part of the story makes me remember him and some fantastic times that we all had. Sorry, let me get back to the story at hand. My friend, the asshole, asked me if I wanted to meet that night for beers so he could make up for the nasty hit. I was a little hesitant but said sure. What time do you want to meet and where?"

"Eight p.m. at the Moose Country Bar and Grill in Mendota Heights, 'he said.' That is where I found myself that night. I met up with him there, and we talked and drank. The one thing he was obsessed with was how my military training went. He mentioned I was a lot bigger than what he remembered. He asked me if I thought I could take him down. I said it doesn't matter, but he kept on insisting that he wanted to find out. So I obliged him and we went outside to the back of the bar by a hill. We fought. Boy, did I want to hurt him so much for his horrible behavior and his arrogance. I beat him down and had a chance to throw him down the hill, and if anyone knows anything about Moose Country and the hill there, they know how dangerous, rough, and high up it is. I had a split-second chance to flip him down the hill or hold him until he gave up. I chose to be the better man and made him concede. When I let him up and shook his hand. I turned around to look down the hill to see what might have happened to him when suddenly I felt a blunt force to my back and found myself falling down the hill, stumbling up and down, hitting the ground, then bouncing up into the air, hitting branches, hearing the shattering sound of breaking glass around me, and finally stopping close to the bottom of the hill, next to an empty keg."

"I was a little shocked, but not surprised by his cowardice. I took a moment to catch my breath and make sure I was intact. My body was hot, my veins in my temple were bulging out, and that anger got me climbing up that hill at an accelerated rate. I reached the top, only to find the coward had left, and that is the only smart thing he did that night, because I would have killed him."

"The next day, I went back to Moose Country to get a better look at the hill that the coward pushed me down. As I look down the hill, seeing all the glass, kegs, broken branches, and rocks, I realized I was very lucky to survive with just a couple of bruises, scratches, and minor cuts."

"One last situation I feel very lucky surviving is one that I felt I had to follow through on. It was one night hanging out with a marine I knew. He had just gotten out of the PTSD program and wanted to go out. I had a friend call me and ask me if I wanted to go out with them. Of course, I said yes. We went to an Irish bar and had many drinks, and flirted with the women in the bar. We were telling jokes and having a great time until this marine said something to a patron at the bar that did not go over very well. Many words were exchanged and some pushing. He was talking with these two girls, and then some guys came over. Maybe they were the boyfriends of these girls, not sure, but the marine I was with was a huge guy, young and ready to fight. I had to get hold of this situation before it got entirely out of control. I was very surprised I could get him out of there in one piece. I talked him into going to my place to chill out, eat something, and have another beer. My friend who was the designated driver got us into his car and safely to my home."

"Once we got there, he seemed somewhat calmer. We ate, had some more beers, and after a while, he started getting angry again and talking crap. He said that he is a God Damn Marine, and nobody can hurt him, and those guys are lucky. I would have killed all of them, he said. At that precise moment, I looked at him, and something inside of me told me he would have not killed them, and all of this was theater for my friend and me."

"I am not saying that at the moment he would have killed none of those patrons, but right now in my apartment. Hours after the incident, saying he would have killed those men. To me, was just more theater. When I laughed, that pissed him off, and he told me he could kill me. I still do not know what propelled me to say what I said to him next, but I did it with no hesitation."

"I looked at him and told him he would not kill me. How could he kill a fellow veteran who had helped him, a friend who was and still is there for him?! He replied, 'I don't care; you are fucking pissing me off. I could kill you and not regret it for a moment.' My reaction was to grab a huge butcher knife from my kitchen drawer and put it in his hand, and I put the pointy end of the knife close to my navel. I said, 'Here, then kill me. You said you could kill me, so do it!' He looks at me like I was crazy, and maybe at that precise moment, I was. I just knew to the bottom of my bones, he would not thrust that knife into me. I can't explain it, but I felt it. In the end, I was correct in my intuition of accessing the situation. He broke down and cried. He apologized to me, and I told my friend to take him home."

"Some people say that it was a very idiotic thing to do. Some others say it was very brave, and others don't know what to say. I can only say that I followed my instinct, my intuition. I was not brave; I was not stupid. I was just trying to help a veteran face himself and realize what he was doing."
"Some might call it luck, some might call it fate. I call it awfully lucky. I might even suggest that I have a guardian angel. Whatever you call it, that luck has flowed over to sports teams. I mean, I love sports, but I never learned why the good luck landed on all those teams as I was working in their states."

"There seems to be no rhyme or reason to why I have had such good luck in my life. Some people attribute it to supernatural interference and/or divine intervention, like having a guardian angel, and others have just called it what it is—I'm just exceedingly very lucky. I don't have a fair interpretation for why my good luck has transformed into fifty-eight national and world championships. The championships first appeared back in 1987 when I first enlisted in the U.S. military and have continued up to the present."

"IF YOUR TEAM IS NOT ON THIS LIST RIGHT NOW, WELL IT COULD BE IN THE FUTURE"

Chapter III
The List

"There are few spheres of life that inspire us to cling to strange superstitions and rituals more than sports. Perhaps it's because sports like football, basketball, hockey, etc. straddle an excruciating line where the outcome of a game seems dictated as much by fortune as it is by plays."
— Bleacher Report

    I asked him if he had a list of the championships, and if I could take a look at it. He asked the server for a pen and paper. She smiled and told him, "Anything for you." As she was retrieving the writing materials, she kept smiling at him. When she returned to our table and was handing him the paper, I noticed she had added her name and number at the top. He smiled at her and then compiled the list as best he could recall it.

    He gave it to me, and as I was reading it, I could not believe all the teams that were listed. It started with the Minnesota Twins, West Germany, Yankees, NJ Devils, and Giants. There were national, international, and college teams on the list. How was this possible? He replied: "I can't explain why it happens or why it has endured for so long; all I know is that it has. I can tell you what the parameters are—what has to happen for the home team to win. That is the only thing I know for sure. I suppose it does not matter how it works but that it works, and if it gets a certain team a national or world championship, does it really matter?"

    I was thinking about it for a while, and he was absolutely correct in making that statement. What I was thinking about right after that was the Vikings winning a championship. "So, I have to ask, what are the parameters for this to work?"

"Well, it started in 1987 with my first substantial job that spring and has continued with every job or assignment that I've had since then. It works like this: For every job or assignment I take; I end up working in that city for at least four to thirteen weeks, and then along comes a national or world championship for the home team." I asked him what he did for work, and he told me he'd had a couple of different positions over the years. It started out, as it does for any typical high school kid, with jobs at local restaurants and fast-food spots. Later he went into the military, and after being discharged, he went to work for a traveling nurse's company. That is what he does today.

As I was reading over the list, I saw that there was a gap in the continuity of championships between 2013 and 2015, so I asked him what this was about. He hesitated and said it was "a little personal to chat about." I paused for a second and then informed him it was imperative to identify the reason for this gap in order to explain the full story about his string of good luck. There was a long, awkward silence before he told me what had taken place during that three-year gap. He looked at me with sad eyes; I could tell because he was the type of individual who wears his emotions on his sleeve. If he was happy, you could see it; if he was angry, you could see it; and if he was sad, you could see it. I could absolutely see that this part of the story made him unhappy to talk about.

"When I quit traveling and working to vacation for a while around New York, New Jersey, and Philadelphia, I realized how very much I despised the area, the corruption, and the lies that people told each other, like that was normal. At one point, the entire area could have fallen into the ocean and I could have cared less. I think it was my distaste for the area that caused the sports teams in that area to experience a very real championship drought. No MLS, MLB, NBA, NFL, or

WNBA team won a championship while I was there from 2013 to 2015. I suppose my sole evidence for that is that there had to be a balance. I have no other explanation for it."

I was speculating in my mind whether this individual had some sort of super-power. There had to be more to the story. I asked him if I could verify his job and rental history to confirm everything he was saying was true. He said, "Sure." He only asked that his name not appear in the article. I agreed. I also advised him we would need to meet one more time after I confirmed everything, which he said was no problem. Collecting my belongings, I stood up and shook his hand. As I was taking off, I swung back around and asked him if he believed in his own good luck. He smirked and said, "My good luck? Hell yes! I am a living testament I have good luck, considering that I escaped death over eight times. But why that transforms into sports teams winning championships, I have no clue. Maybe the universe made me the unofficial god of sports championships." I chuckled. Maybe he was the reincarnation of the Greek god Nike, Hermes, or even Hercules. I was not clear what was going on here, but it was one hell of a story I had to tell. But first I had to ask him one other question before I took off: "Are you working here in Minnesota?" I hoped he was going to say yes so there might be hope for my Vikings, but he told me he was just visiting. I have to say I was more disappointed in that answer than I thought I would have been. So I believed this guy's story. He seemed very authentic and honest, and if his work and rental history checked out, then my Vikings had a chance.

On the next few pages are the lists of the championships that my mysterious friend helped secure. They include FIFA, MLB, MLS, NBA, NCAA, NFL, NHL, and the WNBA. His championships span across all the sport leagues worldwide. It truly is an amazing feat.

YEAR	SPORT	TEAM	TITLES
1987	Baseball	Minnesota Twins	World Champions
1990	Men's FIFA	West Germany	World Champions
1990	Football	Georgia Tech	National Champions
1990	Baseball	Georgia University	National Champions
1991	Woman's FIFA	USA	World Champions
1991	Men's Gymnastics	University of Minnesota	Big Ten Championships
1991	Baseball	Minnesota Twins	World Champions
1992	Baseball	University of Minnesota	Big Ten Championships
1992	Men's Gymnastics	University of Minnesota	Big Ten Championships
1992	Men's Hockey	University of Minnesota	WCHA regular season champion
1993	Men's Basketball	University of Minnesota	NIT Champions
1994	Wrestling	OSU	National Champions
1994	Baseball	Oklahoma Sooners	National Champions
1995	Gymnastics	University of Minnesota	Big Ten Championships
1995	Woman's Soccer	University of Minnesota	Big Ten Championships
1996	Football	Dallas Cowboys	World Champions
1996	Men's Swimming & Diving	Texas Longhorns	National Champions
*1996	Hockey	University of Minnesota	WCHA tournament champion
1997	Woman's Soccer	University of Minnesota	Big Ten Championships
1997	Baseball	Florida Marlins	World Champions

YEAR	SPORT	TEAM	TITLES
1998	Basketball	Chicago Bulls	World Champions
1998	MLS Soccer	Chicago Fire	World Champions
1998	Baseball	NY Yankees	World Champions
1999	Woman's FIFA	USA	World Champions
1999	Baseball	NY Yankees	World Champions
1999	Wrestling	Iowa	National Champions
1999	Basketball	San Antonio Spurs	World Champions
1999	WNBA	Houston Comets	World Champions
2000	WNBA	Houston Comets	World Champions
2000	Baseball	NY Yankees	World Champions
2000	Hockey	NJ Devils	World Champions
2000	Wrestling	Iowa	National Champions
2000	Woman's Softball	Iowa	Big Ten Championships
2000	Woman's Hockey	University of Minnesota	National Champions
2001	Wrestling	University of Minnesota	National Champions
2001	Men's Swimming & Diving	Texas Longhorns	National Champions
2001	Baseball	NY Yankees	World Champions
2002	Baseball	Texas Longhorns	National Champions
2002	Men's Hockey	University of Minnesota	National Champions
2002	Wrestling	University of Minnesota	National Champions

YEAR	SPORT	TEAM	TITLES
2003	Men's Hockey	University of Minnesota	National Champions
2003	Dance Team	University of Minnesota	National Champions
2004	Woman's Hockey	University of Minnesota	National Champions
2004	Dance Team	University of Minnesota	National Champions
2005	Football	Texas Longhorns	National Champions
2007	Woman's FIFA	USA	World Champions
2007	Baseball	Boston Red Sox	World Champions
2008	Football	NY Giants	World Champions
2008	Basketball	Boston Celtics	World Champions
2009	Basketball	LA Lakers	World Champions
2009	Baseball	NY Yankees	World Champions
2010	Woman's Hockey	University of Minnesota	National Champions
2011	Woman's Hockey	University of Minnesota	National Champions
2011	WNBA	Minnesota Lynx	World Champions
2012	Football	NY Giants	World Champions
2015	Woman's FIFA	USA	World Champions
*2016	Hockey	Pittsburgh Penguins	World Champions
*2017	Hockey	Pittsburgh Penguins	World Champions
*2017	Football	Philadelphia Eagles	World Champions
2017	WNBA	Minnesota Lynx	World Champions
2018	Baseball	Oregon State	National Champions
2019	Woman's FIFA	USA	World Champions

* 1996 - Some people say that a tournament champion does not count.

* 2016 - My friend said that because he got married and was getting ready to move that the dark cloud was lifting over the area.

* 2017 - He left and the dark cloud was lifted, so Pennsylvania was the first to enjoy the result, even though he was not working in these states at the time. Remember, there might be more than the 58 championships as these were the only ones he remembered, and we could verify.

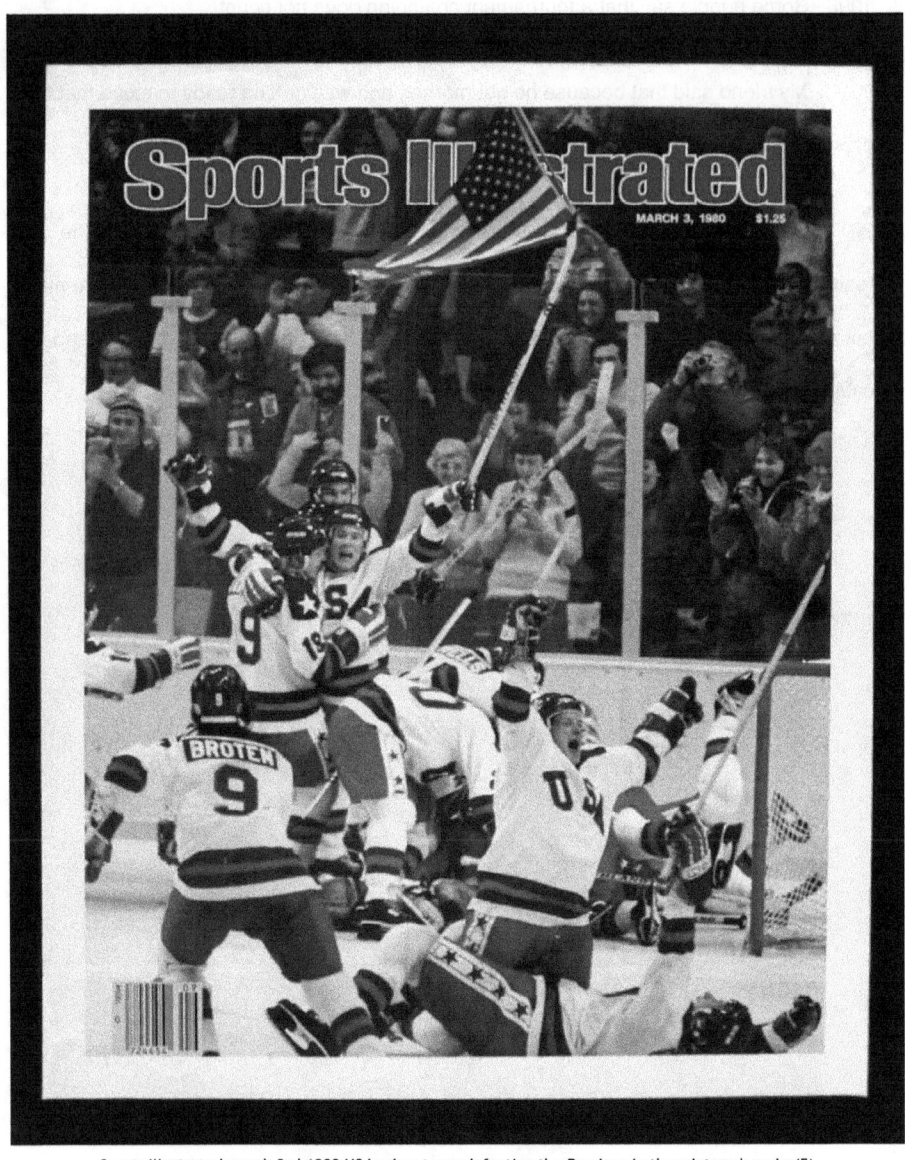

Sports Illustrated march 3rd 1980 US hockey team defeating the Russians in the winter olympics(5)

"LIKE THE 1980 MIRACLE, THIS STORY IS BEING WRITTEN, AND WILL BE REMEMBERED AS ONE OF THE GREATEST MIRACLES IN SPORTS HISTORY"

Chapter IV
The Story

"The hard work puts you where the Good Luck can find you!"
—Anonymous

After running a background check, along with his job and rental history, it all checked out. I was a lot more thrilled than I was when I met him on my birthday. Could this all be genuine? Was I in the presence of a natural, real live good luck charm? It seemed unimaginable, but the facts were right in front of me. I couldn't believe what I was looking at. I hoped he would stay in Minnesota so the Vikings might finally win a Super Bowl. This was at least a remarkable sports story that the world needed to discover. This is how inspirational films come about. I had to get in contact with him to find out more about his life, how this worked, and if he could do it on demand. Could it be possible for him to choose a place to go, and afterward that home team would win a championship? I had to find out.

After calling several times and leaving many voicemails with no response, I was worried he might not be interested in pursuing this anymore with me. But I would not give up. I finally reached him, and he agreed to meet me again. We met at a sports bar near Bloomington to watch the Vikings game. They were playing the Bears, so it was a division rival, and I was ready for an excellent game! I wanted to watch but had to concentrate on finding out more about this miracle worker to see how his magic worked. Again, he was late for our meeting. When he finally showed up, he was wearing a Vikings jersey, and that surprised the hell out of me! But I was glad that he was a fan. It didn't matter either way; I had a job to do. So we sat down and I started the conversation where we left off. He asked me if the information I gave him panned out, and I said, "Yes, it did."

"So then, what else do you wish to know?" he asked. I wanted to know how it worked. Can you do it on demand? Is it magic? Spiritual? Can you drive teams to win with your mind? My questions kept coming, one after another. He interrupted me in mid-sentence and told me to take a deep breath and relax. "Let me tell you what I think about this good luck charm or curse," he said. "It depends on which team you are rooting for." Pausing, I let him talk for the next two hours. I just listened. Here is what he had to say about this whole thing:

"We all love to celebrate when we have something in common, like seeing our favorite team win a championship. Sports seem to do that for the world. Sports figures have fallen into routines, habits, and superstitions that they believe will put them in a better position to win every time."

"Examples for athletes include eating the same thing before every game, wearing college shorts underneath a professional team's uniform, and hockey players pissing on their hands in the belief that they make them softer. For fans, examples are sitting in the same seat, wearing the same jersey at games, and even keeping a rabbit's foot in your pocket during games. These all may seem like strange, inconsequential beliefs, but research reveals that superstitions can literally be linked with improved performance. In short, players gain a psychologically important illusion of control over events that often come down to random bounces here and there. That study (published in the Journal of Applied Social Psychology in 2006) showed that a commitment to rituals is greater for important games, like a league's finals or any playoff game."

"This is what I know to be true. The question that you want answered is why me? Why do all the teams win when I am working and living in that city? I am sure about one thing: A team is always stronger together. What I mean is that having good luck cannot be the only factor to help your team win a championship. You have to believe; the players have to believe; the coaches have to believe. It is always hard, but, 'Stronger together!' And don't forget, there has to be talent on the team."

"Let me tell a more comprehensive story of all the good luck that has played a part in winning these championships, and you decide if it is real or not. Let's start where it all began: The 1987 Minnesota Twins. The Twins, as a team, were bested in virtually every major statistical category in 1987. As ABC play-by-play commentator Al Michaels put it in the pregame show for Game 1, 'They were out of everything!' Insert a little luck, and out comes a world championship."

"Fast-forward to... Wait—I almost skipped the Georgia Bulldogs and their national championship in baseball that year. Read about their season, and it will amaze you they won the title that year. Hey, go Bulldogs! They had a hell of a team that year with many lucky breaks to ensure they got to the championship game. But Oklahoma State was killing teams in the tournament. A lot of sportscasters and people at the tournament said that Oklahoma was the favorite, so they were practically pre-crowned as National champions. But luckily for the Bulldogs, in the championship game, they put on a hell of a defensive show and edged out Oklahoma State for the College World Series title. That was the first and to date their only national title in baseball. Maybe they need me to go back to Georgia. I wouldn't mind; I admire the ladies in Savannah and Tybee Island. The Georgia Tech football team would also celebrate their fourth and, to date, last championship so far."

"My first World Cup was West Germany's championship run in 1990. Germany is always tough and always slightly favored, but the competition was more aggressive that year. It was also the last year that the divided Germanys would compete or win the World Cup."

"The 1991 World Series pitted the American League champion Minnesota Twins (95-67) against the National

League champion Atlanta Braves (94-68). The Twins defeated the Braves four games to three to win the championship. ESPN selected it as the 'Greatest of All Time' in their 'World Series 100th Anniversary' countdown with five of its games determined by a single run, four games decided in the final at-bat, and three games going into extra innings."

"Besides the suspense surrounding the outcome of many of its games, the Series had other highlights. For example, the series-deciding seventh game was a scoreless draw through nine innings and went into extra innings. Minnesota won the seventh game by a score of 1-0 in the tenth inning with their starting pitcher, Jack Morris, pitching all ten innings. Morris was named Most Valuable Player for the Series."

"That Series was also unique because of the standings of the two teams in the prior season. Both finished the 1990 season in last place; before 1991, no league champion had ever finished the previous season in last place, yet this was the case with both the Twins and the Braves. The Twins also won the AL West Division in 1991, with every team in the division having a (.500) or better record. I always found it interesting that when I moved from Georgia to Minnesota, I could see the Atlanta Braves facing off against the Minnesota Twins. My car had Georgia license plates on it, and yet I was living in Minnesota. I think I still had a little sand from Tybee Island in my car."

"Here are some other statistical facts about the Georgia baseball national title you might not know. Georgia got off to a slow start by going 2-3 in its first five games, and then things started to turn around. The Bulldogs won eighteen of their next nineteen outings. In that stretch, Georgia pieced together a fourteen-game win streak. Opposing the Bulldogs were the Oklahoma State Cowboys, a team that had already run roughshod over the entire CWS field. Coach Webber took a calculated risk and started freshman lefty Stan Payne against the heavy-hitting Cowboys. Holding true to form, Payne continued Georgia's solid pitching. With his teammates staking him to a 2-0 lead through five innings, Payne allowed only an infield single until OSU finally scored in the sixth to cut the lead in half at 2-1. Payne then yielded to Fleming in the seventh inning. Fleming retired eight of the ten men he faced and struck out the side in the ninth inning to preserve the Georgia win."

"When all was said and done, the Bulldogs produced the most successful year in their history. They finished with a (52-19) record. In the process, Webber was named NCAA Coach of the Year. Fleming, Payne, shortstop J.R. Showalter, designated hitter Brian Jester, and outfielder Ray Suplee were all named to various All-American teams. Do you think that luck didn't play any role in this successful season? Of course there was."

"Now we have to move on to another outstanding team hailing from the great state of Oklahoma—the University of Oklahoma Sooners. In 1994, the Sooners took home the national title, capping off an incredible NCAA Tournament run. The Sooners then went undefeated in both the regional tournament in Austin, Texas and the eight-team, double-elimination World Series. At the time, OU became only the fifth team since the College World Series had moved to Omaha in 1950 to accomplish such a feat."

"To further show the Sooners' dominance during the postseason, Oklahoma trailed in just one of the seventy-two innings in the NCAA Tournament. OU finished the series as the top hitting (.327) and pitching (2.37 ERA) team while ranking second in defense, committing only five errors."

"Leading the way for the Sooners was senior centerfielder Chip Glass. He was named Most Outstanding Player of the series. Glass, who hit only three home runs during the regular season, nailed three round-trippers during the Series, including one in the championship game. Over the four-game tournament, the Ukiah, California native batted (.389) with four RBIs and three stolen bases."

"With the game tied at two after three innings, Oklahoma jumped ahead and claimed a commanding lead that it would not relinquish. Aided by four hits and three Georgia Tech Yellow Jackets errors, the Sooners scored five runs in the fourth inning to break open a 7-2 lead. Two frames later, a two-run double by Georgia Tech's designated hitter Michael Sorrow in the top of the sixth brought the Jackets back to within three runs before a three-run homer by Sooners sophomore designated hitter Damon Minor capped off a four-run bottom of the sixth, putting the game away for the Sooners for good."

"Junior left-fielder Aric Thomas and senior right-fielder Darvin Traylor led all Oklahoma hitters with three hits and runs apiece; both drove in one run each. Junior right-hander Tim Walton pitched five scoreless innings in relief a week before against Arizona State in the Sooners' second game of the series. He picked up the win for OU after throwing 2.1 strong innings in relief of senior starting pitcher Kevin Lovingier. Oklahoma's ace closer, junior right-hander 'Bucky' Buckles, came on for the final 3.2 innings of work to close out the game and record his school and Big Eight record fourteenth save of the season."

"Buckles' performance earned him a spot on the all-tournament team along with five other Sooners: Glass, senior second-baseman Ricky Gutierrez, sophomore first- baseman Ryan Minor, Traylor, and junior pitcher Mark Redman. As a team, Oklahoma set a championship game record with its thirteen runs and tied the record of sixteen hits. They did not have any superstars. Their coaching was okay, nothing spectacular, but combine those two with a little luck, and you have got yourself a championship team."

"In 1993, the Oklahoma State wrestling program was hit by NCAA sanctions that had a significant impact on the program. The Cowboys were coming off of a second- place finish in the 1992 NCAA Championships. They imposed penalties on OSU Wrestling for violating NCAA policies. Some stipulations included three years of probation, loss of competition on television, and most notably, a ban from the 1993 NCAA tournament. Guess what? They still won the national championship the next year in 1994."

"The 1996 Dallas Cowboys season was the franchise's thirty-seventh season in the NFL and the third year under head coach Barry Switzer. Following their victory in Super Bowl XXX, the Cowboys endured a harsh and wild year, failing to improve on their (12-4) record from 1995 but still reaching the playoffs with a (10-6) record. Star receiver Michael Irvin was suspended by the league for the first five games and, before the playoffs, was accused along with lineman Erik Williams of sexual assault."

"Another controversy came up when writer Skip Bayless published a scathing account of the Cowboys' 1995 season. Longtime trainer Mike Woicik also left the team after the season following a sideline dispute with coach Barry Switzer.Against the Chicago Bears in week one, running back Emmitt Smith would leave the game late with an injury that left him temporarily paralyzed. Though not career-threatening, Smith's injury would hamper his effectiveness for the duration of the season. Adding to their difficulties, they suspended Irvin for the first five games because of his highly publicized off-season incidents. Star cornerback Deion Sanders would be the first player in the NFL's modern era to start several games on both offense and defense. Charles Haley, a major defensive force for the Cowboys during the prior four seasons, missed most of the 1996 season with an injury. Tight end Jay Novacek, a key offensive threat in recent seasons, missed the entire season because of an injury."

"After losing three of their first four games, the team would return to form, winning three straight before defeating former head coach Jimmy Johnson, then head coach of the Miami Dolphins, and were on the way to their fifth consecutive NFC East title. Although Dallas still moved the ball well on offense, they had serious late-season trouble scoring touchdowns, finishing just twenty-fifth in the league in points scored. They won games against Green Bay (21-6) and New

England (12-6, the next season's eventual Super Bowl participants) without scoring a touchdown (seven field goals against Green Bay and four against New England); Dallas' defense finished third in the league in fewest points allowed."

"The game of the year came on November 10 at San Francisco. With the Cowboys standing at 5-4 and the Niners at 7-2, the Niners had won three straight meetings with the Cowboys since they hired Barry Switzer as head coach. Sacks by Broderick Thomas and Jim Schwantz knocked Niners quarterback Steve Young out of the game and Elvis Grbac, who'd shredded the Cowboys defense the previous year, came on. After the Niners blew a 10-0 lead, they took a 17-10 lead with 11:30 left in the fourth quarter. The Cowboys drove to the Niners red zone, but Marquez Pope picked Troy Aikman off in the end zone with 6:30 to go. Grbac, however, was intercepted at his six- yard line on the ensuing play by Fred Strickland, and three plays later, Aikman connected with Eric Bjornson for the tying touchdown. In overtime, a big Emmitt Smith run set up Chris Boniol's winning field goal. Combined with a Redskins loss to Arizona, the Cowboys win allowed them to win the NFC East."

"The Cowboys solidified their status as one of the most prolific dynasties in NFL history by becoming the first of now only two teams to win three Super Bowls in four seasons, knocking off the Pittsburgh Steelers, 27-17, in Super Bowl XXX thanks to a very unlikely hero—a concept called 'a little good luck.'

"In that Super Bowl, cornerback Larry Brown, lost in the shadow cast by fellow cornerback Deion Sanders, intercepted two overthrown passes from Steelers quarterback Neil O'Donnell. That sealed the victory and earned him Most Valuable Player honors at Sun Devil Stadium in Tempe, Arizona, on January 28, 1996. Holding a 20-7 lead going into the fourth quarter, the Cowboys endured an intense Pittsburgh rally. The Steelers scored ten consecutive points to close the gap to 20-17 with 6:36 left to play. The Steelers held, got the ball back again two minutes later, and took their first lead of the game. But Brown ended any potential comeback, intercepting his second pass of the ballgame and returning the ball to the Steelers' six-yard line, from which Cowboys running back Emmitt Smith clinched the Super Bowl title with a four-yard touchdown run."

"This emotional finish, which left the Cowboys players and coaches drained, was sure not indicative of how this game began before 76,347 people in the desert sun."

"Now, 1997 was an exceptional year for me. I can't get into details, but I have very fond memories from 1997 to 2001, some of my best years. I felt good, I felt alive, I loved life."

"So let's continue on to the 1997 World Series, the ninety-third edition of Major League Baseball's championship series, which began on October 18 and ended on October 27 (after midnight on October 26). It featured the Cleveland Indians, playing in their second World Series in three years, and the Florida Marlins, which had set a record by reaching the World Series in only their fifth season. The Marlins, who were underdogs, capped a stunning season. They defeated the Indians four games to three to win their first World Series, making them the first wild-card team to ever win the Series. The Game 7 finale was decided in extra innings on a Édgar Rentería single."

"Soon after they completed game 7, rumors on the internet started to spread that in 1989, four years before the Florida Marlins made their debut, their 1997 World Series victory was predicted in Back to the Future Part II. In reality, the movie predicted that in 2015, a Miami team with an alligator mascot would lose to the Chicago Cubs in the World Series. Ironically, the Cubs' actual 2015 season also ended in late October and in a four-game sweep. This time they lost to the New York Mets in that year's NLCS. It is funny how people will acknowledge that a movie can predict a championship or somehow influence the outcome. I say, hell yes it can! Just like every time I take an assignment, a national or world championship for some home sports team comes with it."

"Next stop: Chicago. That was a fun place to be when the Bulls won their sixth championship. Remember when Jordan put up his fingers after one championship, maybe two, maybe three championships with that Jordan-like grin? Come on, we all remember that. Well, he should have held up six fingers instead. Here are some facts:

"The 1998 NBA Playoffs concluded with the Eastern Conference champion Chicago Bulls defeating the Western Conference champion, the Utah Jazz, four games to two in the NBA Finals. The Bulls also achieved a second 'three-peat,' unrivaled since the Boston Celtics did it in 1966. The 1998 playoffs were the last involving the Jordan-led Bulls; Chicago didn't make the playoffs again until 2005. They named Michael Jordan NBA Finals MVP for the sixth and final time."

"The 1998 Finals garnered the highest Nielsen TV ratings in NBA history at 18.7. This surpassed the Nielsen ratings for the 1998 World Series, marking the first time the NBA had a higher rating in its championship round than any of Major League Baseball's championship round. After Jordan made a layup to make it 86-85, the Bulls needed to stop the Jazz from scoring again. When John Stockton passed the ball to Karl Malone, Jordan stole the ball away and dribbled down the court. Guarding him was Bryon Russell, one of the Jazz's best defenders. With ten seconds remaining, Jordan dribbled right then and crossed over to his left. Jordan pushed Russell off with his free hand and hit the 20-foot jumper to give the Bulls an 87-86 lead with 5.2 seconds left. After a time-out, Stockton missed a potential game-winning three-pointer, giving the Bulls their sixth NBA title in eight years. Jordan, who scored forty-five points and whose game-winning shot has been immortalized around the world, was once again named Finals MVP."

"To date, the series remains the last Finals appearances for both the Bulls and Jazz. After the season, the Bulls dynasty broke up. The Bulls would not make the post-season again until 2005, win a playoff series until 2007, or earn the Eastern Conference top seed until 2011. I have not been back to Chicago since 1998; coincidentally, neither has an NBA championship."

"We head down south now to the big state of Texas by way of New York, where the Yankees won the World Series in 1998 and 1999. Off to Texas I went for another assignment, where the 2000 WNBA Championship played out."

"The Houston Comets were the second-seeded champions of the Western Conference. They defeated the New York Liberty, first-seeded champions of the Eastern Conference, two games to none in a best-of-three series. This was Houston's fourth title. Don't forget, the Yankees also won that year."

"There were several college championships from 2000 to 2005; look for yourself. You have the chart to see the teams and what years they won. Now fast-forward to 2005, and the Spurs win a world championship. I was there on another assignment. I love Texas. Love the hospitality and the beautiful women."

"The 2005 NBA Finals was between the Western Conference champion San Antonio Spurs and the Eastern Conference champion Detroit Pistons for the title, with the Spurs holding home-court advantage and the Pistons as defending champions. They played the series under a best-of-seven format. (Unlike the previous three rounds, the team with home-court advantage hosted games one, two, six, and seven if all were necessary.) It marked the Pistons' first NBA Finals loss to a team other than the Laker's since 1988. The Spurs won the series four games to three in the first NBA Finals to go to a Game Seven since 1994."

"This series was not only the first matchup of the previous two NBA champions since the 1987 Finals (Laker's defeating Celtics, 4-2), it was a matchup of the two premier defensive teams of that era. From the 2002-03 season to the 2004-05 season, the Spurs and Pistons finished in the top three in least points allowed. In 2003, Detroit was number one and San Antonio number three. 2004, the teams finished in a tie for the number-one spot. In 2005, the Spurs were number one, and the Pistons were number three. They considered the Spurs more capable of playing at a faster pace, as evidenced by their convincing win against the high-scoring Phoenix Suns during the Western Conference Finals. However, both teams performed well when they scored over 100 points (Detroit was 22-3, San Antonio was 28-2). Sportswriters across the country considered this to be one of the few too-close-to-call NBA Finals to be held. Most picked the series to go to six or seven games."

"Okay, let's move to Pittsburgh and see what they did when I was there. With the win, the Steelers joined the San Francisco 49ers and the Dallas Cowboys as the only franchises too have won five Super Bowls. The Steelers' victory was their first Super Bowl victory since Super Bowl XIV. Pittsburgh, which finished the regular season with an (11-5) record, also became only the fourth wild-card team, the third in nine years, and the first-ever number-six seed in the NFL Playoffs to win a Super Bowl. The Seahawks, in their thirtieth season, were making their first-ever Super Bowl appearance after posting an NFC-best (13-3) regular-season record."

"Let's move on to one of my favorite Super Bowls. I regard this game as one of the biggest upsets in the history of professional sports. It was one of the greatest Super Bowls. The Patriots entered the game as twelve-point favorites after becoming the first team to complete a perfect regular season since the 1972 Miami Dolphins—and the only team since the league expanded to a sixteen-game regular-season schedule in 1978. The Giants, who finished the regular season with a 10-6 record, were seeking to become the first NFC wild-card team to win a Super Bowl. They were also looking for their third Super Bowl victory and first since they won Super Bowl XXV seventeen years earlier."

"This Super Bowl was also a rematch of the last game of the regular season in which New England won, 38-35. The game is best remembered for the Giants' fourth-quarter, game-winning drive. Down 14-10, New York got the ball on their own 17-yard- line with 2:39 left and marched eighty-three yards down the field. In the drive's most memorable play, David Tyree made the "helmet catch," a leaping, one-handed catch, pinning the football with his right hand to the crown of his helmet for a 32-yard gain."

"The telecast of the game on Fox broke the then-record for the most-watched Super Bowl in history with an average of 97.5 million viewers nationwide. Part of the reason for the interest was that the teams were from rival cities between New York City and Boston, which are only about four-and-a-half hours apart by car. The Super Bowl echoed the fierce rivalry between the New York Yankees and the Boston Red Sox, games that are often televised by Fox."

"The teams' fans wore Yankees and Red Sox hats and showed off Fenway Park and Mickey Mantle T-shirts. Patriots' fans said that they had no natural hatred for the Giants. Many old-time New Englanders, in fact, grew up rooting for the Giants before Boston got its AFL franchise in 1960. They were more accustomed to rooting against the Jets. Giants fans, however, harbored a great hatred for the Dallas Cowboys, Philadelphia Eagles, and any team from the Boston area. Many Giants fans also wore the hats and shirts of the New York Mets and the New York Rangers to show they're hatred for the New England Patriots."

"Before the game, 'Patriots' fans said they wanted to continue the trademark 'Yankees Suck!' chant, which began after they won Super Bowl XXXVI in 2002 (which followed the Yankees' loss in the 2001 World Series). Giants fans wanted revenge for 2004 when the Red Sox came back from a 3-0 deficit to beat the Yankees in the ALCS en route to winning the World Series. The post-game celebrations even played up the rivalry, but to a lesser extent. Giants fans called it revenge for the Red Sox coming back from 3-0 down to beat the

Yankees in the 2004 American League Championship en route to ending the 'Curse of the Bambino.' In response to Patriots fans chanting "Yankees suck!" while celebrating their Super Bowl XXXVI victory, Dan Shaughnessy of The Boston Globe wrote, 'Can you imagine a Giants or Jets celebration in New York City in which a New York player would take the time to chant, 'Red Sox suck?" He was referring to retaliation for the 'Yankees suck!' chants."

"After the game, Giants fans chanted '18 and 1,' reminiscent of the infamous '1918!' chant that Yankee fans yelled at the Red Sox each time they visited Yankee Stadium until the Red Sox won the 2004 World Series. The Giants' fans directed the same chant towards Patriots' fans as they left the stadium, referring to the Patriots' final record. Giants players and fans also heard the same chant during rallies and the celebratory parade. There are a lot of football fans who believe luck is the only thing that got the Giants past the unbeatable New England Patriots that year. We will never, ever forget the famous 'helmet catch.'

"Love or hate the New York/Boston rivalry, it made for so many good sports stories, and 2008 basketball did not disappoint. The Giants took away a perfect season from the Patriots, and the Celtics come back and won a national championship. They were due, and they did it in a Bostonian way—as scrappers, fighters. Along with what the Irish like to say was a little luck."

"The 2008 title was Boston's first since 1986 and seventeenth overall. It marked the first time since 2000 that the top seeds from both conferences met in the Finals and the first time since 2003 that any top-seeded team played in the NBA Finals. With the Laker's appearing in the Finals for the first time since 2004 and a record twenty- ninth time overall. The Celtics appeared in the Finals for the first time since 1987 and a second-best twentieth time overall. The Celtics were making their first NBA Finals appearance since a six-game loss to the Laker's in 1987. Over the next twenty years, the

Celtics would suffer through several lean years, not making the playoffs in nine of those years."

"Following the departures of Larry Bird and Kevin McHale, the Celtics suffered through several tragedies including the passing of Reggie Lewis in 1993, a franchise-worst 15-win season in 1996-97, the ill-fated hiring of head coach Rick Pitino, and the deaths of franchise patriarch 'Red' Auerbach and former player Dennis Johnson. The miseries culminated in a 24-win season in 2006-07, which was highlighted—or should I say 'low-lighted'—by a franchise-worst 18 straight losses. But hope was on the horizon. Luck plays a huge part in sports and in life. Ask any combat soldier if they would rather be just okay at their job or lucky while on the job."

"The next year, 2009, was okay. I was not a fan of California. Every physical fight I had ever been in was with some guy from California, so my love for the state was not as strong as it should have been. Nonetheless, I took an assignment there, and the Laker's won a championship. It was not flashy, it was not remarkable, and nobody remembers it. I wish the Laker's had stayed in Minneapolis."

"Let's move on to 2012 because that is where it all ended, and in New York of all places. This would be my last year of taking assignments, so it was only fitting that my last championship should come at the expense of one of the greatest teams in football history and one of the greatest curses of all time. With a subpar team and my good luck, the Giants snatched victory from New England once again, and in superb fashion, I might add. Besides winning their fourth Super Bowl in team history, the Giants set a record for the worst regular-season record (9-7, with a winning percentage of 56.3) by a Super Bowl champion. The Patriots entered the game with a (13-3) regular- season record and were also seeking their fourth Super Bowl win."

"The game was also a rematch of Super Bowl XLII, which New York also won, 17-14, to spoil New England's run at a perfect 2007 season in what I now consider to be one of the greatest upsets in sports history."

"The broadcast of the game on NBC broke the record at the time as the most-watched program in American television history, set during the previous year's Super Bowl. An estimated average audience of 111.3 million U.S. viewers and a total audience of 166.8 million, according to Nielsen, watched Super Bowl XLVI. That meant that over half of the American population watched at least some of the broadcast. The game also set the record for most tweets per second during a sporting event, with 13.7 million tweets from 3 to 8 p.m. (PST). The Giants became the first team to have won Super Bowl games broadcast on all four U.S. national networks: CBS, ABC FOX, and now NBC."

I had to take a break, as that was a lot of information to take in. After I excused myself from the table to take a five-minute stretch, I was thinking of all those teams, games, and championships. It was quite a remarkable resume. I went back to the table to sit down and see if I could get some more information from this guy. I asked him what the next chapter in this story might be. He replied that this is where the story ended because after that his life took a turn he never saw coming.

"I suppose that is why the championships stopped. Maybe the reason I was a good luck charm was because of how I felt about my assignments. I wanted to see the underdog win. I am not really sure. What if I could play a part in your team winning a world championship? That would be a day of celebration! I think if you are a fan, player, or coach, you would do anything to maximize the chances of your team winning that national or world title. I want to see if I can reach fifty-nine national and/or world championships. Don't you?"

"I want you to come and work in Minnesota so my Vikings can finally win a Super Bowl," I replied.

He started laughing. "Yeah, I am sure—you and every other sports fan."

"When did you realize that this 'good luck' and all those national championships followed you?"

"I did not notice until my brother brought it up one day when we were watching the Vikings play on TV. One of the announcers was talking about the role of luck in the "'Giants' Super Bowl win over the Patriots, and my brother said, 'Hey, you were in New York that season, right?' I said, 'Yeah, I believe so.' Then he asked me about my assignments while scanning the internet, and what did he find out? Exactly what I am telling you. We joked about it, and he gave me this stupid nickname that only my family calls me."

"What was the nickname?"
"My brother called me the M&M Man."

I laughed out loud. The first thing that came to my mind was the candy. I had an image of one of those M&M candy characters in a Vikings jersey. "I have to ask you, what does the M&M stand for?"

"The Minnesota Midas Man. You get it? Everything I touch turns to gold, or in this case, every place I go turns the home team into a national or world champion."

I liked his nickname. It was appropriate in this case. But what I remember about one of the many legends about Midas was that he had this great gift given to him by the gods, but after a while he regretted it, especially when he accidentally turned his daughter to gold. "So, after your brother opened your eyes to the facts surrounding your good luck, did you do any research into this phenomenon?" I asked.

"I did a little research, and I found out a lot about sports and good luck, and how they have a symbiotic relationship. It would surprise you how many stories about athletes and their superstitions or fans and their superstitions affect certain outcomes. It even applies to corporate behavior. The superstitious stuff that humans do has got to be one of the funniest but most endearing parts of humanity."

"LUCK WILL BE THERE WHEN YOU NEED IT, IF ALL THE PIECES ARE IN PLACE"

Chapter V
Good Luck

"Realizing full well that fine condition and confidence will not in themselves make a champion, it is my belief, however, that they are essential factors."
— Major Taylor

"There are many other reasons why having a good luck charm for your team is the way to go," the M&M Man told me as our conversation continued. "It works, and players believe in them; fans believe in them. If you carried around a good luck charm as a kid, you might have noticed that you sometimes performed better on tests or scored more goals at soccer practice. This extra fortune may have just been in your head, but psychologists say that doesn't matter. In fact, that's precisely why your lucky charm worked. Science backs this phenomenon."

"Take, for instance, a 2010 study in which German psychologists challenged twenty-eight university students to a game of mini-golf. Participants who were informed they had a 'lucky' ball ended up performing thirty-five percent better than the subjects who were told they were using an ordinary ball. The same researchers then did a follow-up study of forty students who admitted to having lucky charms. They asked them to take a memory test either with or without the charms. Those who kept their totems had higher confidence and scored better on the test."

"Deep down, athletes understand that certain actions don't really affect the outcome of a game. But once the idea that these actions might affect their performance is inside their heads, they may choose to do them anyway because there's little down-side. I have fifty-eight national and world championships under my belt. There is no downside to having me as your team's good luck charm. I have my ritual for taking a work assignment and a routine to make sure that the home team wins their world championship. But that also meant that bad luck befell the teams they played. There are many examples of that."

Here is a story that Charles Bricker, a staff writer for the [Fort Lauderdale] Sun- Sentinel, wrote:

It was the red towel that did it, New York Giants coach Bill Parcells revealed Monday.

That ratty, beat-up good-luck hand cloth first sent him by New Jersey high school coach Rich Conti four years ago.

"The towel has never lost," the superstitious Parcells explained at a news conference the day after the Giants held on to defeat Buffalo 20-19 in Super Bowl XXV.

"I've got it in my bag right now and, Rich," he told reporters, as if they would somehow transmit the message, "it will be back in the mail this week. Back in Warwick, N.Y., Conti was bubbling over after the Giants win."

"That towel is two-for-two for me in championship games, and now it's two-for- two for coach Parcells," he said in a telephone interview from his home.
Conti had sent the towel to Parcells toward the end of the Giants' first Super Bowl season in 1986. Now, after four years, he figured it was time to act again.

"The towel has been on a shelf in my office and on this particular year, I don't know, it just felt like the Giants could use it."

Two days before the Giants left for San Francisco to play the 49ers in the NFC championship game on January 20, Conti told himself: "Geez, I'd better get this back into Parcells' hands." He gave it to his friend, Giants film producer Tony

Ceglio. "Don't forget to give this to him," Conti admonished his buddy. Parcells took it to San Francisco, where his team beat the 49ers 15-13, then packed it for Tampa for the Super Bowl.

The Giants coach has written Conti, thanking him for the good luck towel, but the two men haven't talked since they met at a football camp in 1983.

Conti inherited the towel in 1985 from another high school coach at New Milford High School in New Milford, N.J. "He told me he won the 1983 state championship with it, and it might work for me.

"I used the towel in 1985 and 1986 and won the North Jersey Section I state championships both years," said Conti. His teams were 9-2 and 9-1-1. While his team was winning its second straight title, the Giants were kicking into high gear near the end of the 1986 season and seemed headed for the playoffs.

"I wrote a note to Parcells, telling him, 'Coach, even though we all know block- ing and tackling wins games, a little luck doesn't hurt.' I told him the story of the towel and hoped he could use it. I knew how superstitious he was. After he beat the Broncos in Super Bowl XXI, he sent the towel back in a manila envelope with a friendly note." The towel is well worn. "In fact, it's kind of ratty looking," said Conti. "I don't know if you'd want to use it on your face."

Conti watched the Super Bowl at a friend's house. When Buffalo kicker Scott Norwood lined up for a 47-yard field goal with four seconds to play, a kick that could have won the game, Conti remembers telling himself: "He'd better miss or the jinx is off the towel."

If the towel has such magical powers, why didn't Conti send it to Parcells in the years the Giants didn't go to the Super Bowl?

"I don't know. As a coach, you have a feeling about when the time is right. You have to pick your spots," Conti explained.

Conti, 43, grew up in New Jersey and has been coaching high school football there for 21 years. He teaches history at Emerson High in Emerson, N.J., and coaches at Northern Valley Demarest in Demarest, New Jersey

As far as he knows, neither he nor Parcells has ever washed the towel. "To be honest with you," Conti said, "the thing is so old and worn out, you couldn't tell the difference if it was. But I'm not going to complain about whether it's dirty or clean as long as Giants know luck plays a role in getting to the Super Bowl.

Here is another story about how luck plays a role, written by Michael David Smith on February 5, 2012:

Giants G.M. Jerry Reese pointed out last week what a thin line there is between a successful season and a failure, noting that last year the Giants won 10 games and failed because they didn't make the playoffs, while this year the Giants are in the Super Bowl after winning only nine games in the regular season.

The difference between a great year and a failure can be the difference between good luck and bad luck in a couple of crucial games. And the Giants know that this year, they've had some lucky breaks.

"Winning is about a good game plan and execution," Giants center David Baas told the New York Times. "But it's also a little bit about luck."

The New York Times detailed five distinct moments this season when lady luck gave the Giants a gift, close games that went the Giants' way but could have gone the other way with a different call or a different bounce of the ball. If the

Giants take the Vince Lombardi Trophy tonight, that good luck will have played a part in it.

Of course, the Giants have also had some bad luck. For instance, in both of their games with the Packers—a regular-season loss and a postseason wins with highly questionable calls that went against them. It's not like the Giants have gotten all the breaks. But sometimes, getting a few of the breaks is the difference between being in the Super Bowl and missing the playoffs. That was the case for the Giants this year.

Look, my new friend has said many times that good luck is just part of the winning equation. The Giants' 2012 season is the epitome of that statement. Let's take a quick look at some times where luck played a part in that winning season. I think the cutest story is about a little teddy bear, subsequently called "Little Teddy Bear." It was there in 2008, along with my friend, and Little Bear along with my friend did it again in 2012.

The 2012 season was a seesaw season between losses because of some terrible calls, a lot of injuries, and some extreme luck going their way during the season. Some examples of this are:

1.) The fumble that wasn't. Week 4: Oct. 2 against the Arizona Cardinals.
2.) The pass-interference call. Week 8: Oct. 30 against the Miami Dolphins.
3.) Lost in the lights. Week 14: Dec. 11 against the Dallas Cowboys.
4.) The last-second heave. Division Playoff Round: Jan. 15 against the Green Bay Packers.
5.) The muffed punt. NFC championship game: Jan. 22 against the S.F. 49ers

The New York Giants should have lost every one of these games if it was not for luck somehow intervening and turning the tide in their favor. Let's not forget that my friend was also part of this lucky season. Jerry Reese is quoted as saying how things outside of a team's control can often make all the difference in determining public perception. Or even a season.

"Last year we win 10 games and we don't qualify for the tournament, and you're not that smart," Reese said. "This time, we win nine games, win a division, less games, and now it seems I'm pretty smart again. It just comes with the territory, and that's just part of it. It just is what it is." This happened to be a year that nine wins was enough for the Giants to get in the playoffs. That wouldn't work most years.

"That's just how it goes. You need to be very good to make the Super Bowl, and you usually need to be a little lucky. You can go to the New York Times to read more about all these games." Michael David Smith of the NYT wrote the story. The next section is about how bad luck, or curses, play a part in sports.

"The weirdest thing about the 'Madden Curse' is that it seems to exist. Rationally, we all know that there's no real correlation between appearing on the Madden video-game cover and the cataclysmic collapse of so many once-promising NFL careers. But there is a lot of anecdotal evidence to support this theory. Here are just a few players who have fallen victim to the dreaded curse:"

"How about Daunte Culpepper? The season after appearing on the cover of Madden 2002, the Vikings quarterback went 5-11 and broke the record for the most fumbles in a season. He blew out his knees in 2005 and never returned to form."

"Or Michael Vick? After appearing on the cover of Madden 2004, the Falcons quarterback broke his leg in a preseason game, which kept him out all season. It was the same year his dog fighting ring came to light."

"How about Donovan McNabb? The Eagles quarterback landed on the cover of Madden 2006 and incurred a sports hernia on the first game of the next season. McNabb played hurt all season until a torn ACL mercifully ended his year early."

"Then there's Shaun Alexander. After his MVP season in 2006, the Seahawks running back landed on the cover of Madden 2007. He suffered an early season injury that caused a sharp decline in his production—and he was out of football just over a year later."

"I would not be doing my job if I did not mention the 2008, 2009, and 2010 football players who were also on the game cover."

"In 2008, it was Brett Favre. The Green Bay legend had gone to the Jets that year and had a dismal year, but rebounded with the Vikings a year later only to lose a heartbreaker to the New Orleans Saints."

"On to the 2009 Madden cover. That was Vince Young's year. Is he even around anymore? He ended up being released the following year by the Tennessee Titans. Sorry, man."

"In 2010, the Madden game folks did something they had not done before. They put two players on the cover. I will say this: They played it pretty smart. Two players who were at the peak of their careers. I can't really say that anything bad happened to Troy Polamalu or Larry Fitzgerald."

"But Peyton Hillis? An inexplicable choice to begin with, the Browns running back somehow landed on the cover of Madden 2011 after a flash-in-the-pan season in Cleveland. He's done exactly what since then?"

"Now I'm going to hand you a list of sports curses that have plagued some of our most beloved teams for quite some time, starting with the long-term suffering Chicago Cubs fans." This appeared on the list he handed me:"

1. The curse of the billy goat stems from a local bar owner who was booted out of Wrigley Field during the 1945 World Series because people complained about the smell of his companion, a goat. He and the goat left, but not before putting a curse on the team.

2. The curse of the black cat: Locked in a tight division race, the Cubs met the Mets in September 1969 for two critical games. Early in the game, a black cat mysteriously appeared from the stands at Shea Stadium and sauntered past the Cubs dugout. The Cubs lost the game and later the division. The Mets won the World Series.

3. The curse of Steve Bartman: Steve Bartman, as you may recall, is one of many Cubs fans who attempted to catch a foul ball in Game 6 of the 2003 NLCS. Fan interference prevented a potential Cubs catch, which would have put them four outs away from winning. The Cubs somehow lost the game and then the series, and all of Chicago blamed Bartman. As well, they should.

As I read the list, he took a break in order to get a beer, and I excused myself to take a leak. I looked at my watch, and it was well over ninety minutes so far. I was so immersed in his story that the time had flown by and I had to figure out how this all worked. When I got back to the table, there was a beer waiting for me. I thought I should not drink during the interview, but it felt appropriate for a sports story, especially one like this. I accepted the drink with one condition: He would have to prove to me somehow that this works—that this magic, this good luck, really worked and continued to work for him.

He smiled and told me, okay, no problem. He asked if he could continue with his talking points. I nodded politely and said "of course" in a slightly quieter voice. He then continued to explain why good luck happens.

"There was one experiment that caught my eye. One of psychology researcher John Wiseman's most renowned experiments involved a newspaper and a set task: Count how many photographs appear inside. The trick, however, was that on the second page of the newspaper, Wiseman had planted a large advertisement that read, 'Stop counting.' There were forty-three photographs in this newspaper. Of his 400 participants, those who regularly said they experienced good luck identified and read the 'stop counting' message within two seconds; those who felt they were unlucky didn't see it and continued to count the photographs."

"Wiseman pointed out that this wasn't a matter of stupidity but of focus. Open- ness is not just a social capability but an approach to tasks. 'Unlucky people miss chance opportunities because they are too focused on looking for something else,' he explained. Being too task-oriented, he concluded, could sometimes be a disadvantage because it distracts from other opportunities that arise along the way."

"There are other explanations that scientists and others have given for luck. These include telekinesis, moving things with your mind, and putting thoughts into people's heads. I don't trust in that stuff, but the government has put time and money into researching these types of things So, if you do not believe in that sort of thing then you should ask the United States government for your taxes back. It was your taxes and mine that funded that type of research."

"Maybe I will help teams win championships with my mind. Maybe not. Lots of people believe in things that they cannot see or touch or even prove too exist. But they continue to believe in them anyway, and not being able to prove their existence does not make them any less important in their minds. Nor does it make these things any less believable to them. You have proof of my long history of bringing national and world championships to the home teams where I have worked and lived for over thirty-two years. You also have the law of averages to consider."

"According to the law of averages, a sample's behavior must line up with the expected value based on population statistics. For example, suppose I flip a coin 100 times. Using the law of averages, one might predict that it will be fifty heads and fifty tails. While this is the single most likely outcome, there is only an eight percent chance of this occurring."

"So, if we are to conclude the likeliest outcome is about one out of ten, then my average is way above that. Not counting the gap of a few years, I have a 58-straight win streak, so how do you explain that? I am just an average man with a working man's mind. I suppose there are mathematical geniuses out there who can explain it better to you. I just don't really have an explanation for how or why it works, only a where and when."

"I already told you this the night of your birthday about the Yankees. When the Steinbrenner family dug up the foundation of the brand-new Yankee Stadium, they found out one of the construction workers embedded a Red Sox jersey in the cement of one of the construction sites. It cost the Yankees a lot of money to dig up the foundation just to prevent a jersey from their rivals from becoming a permanent stain on the team. When you support something like that, why is it hard to believe in this?"

I sat there for a couple of minutes after he finished talking, trying to get a grip on all of this in my mind. I was thinking about the universe and how it has positives and negatives, and it all balances out. Then it hit me: If he had all this good luck, then where did all the bad luck go? So, I asked him: "I believe all you're good luck, but where was the balance? There had to be bad luck around you, too. So, what was that bad luck? Where and when did it happen, and who did it happen to?" He paused for a time and replied with a smirk. "That is the question I have been waiting for you to ask.

You are absolutely right about the bad luck and the balance that is needed for the good luck to happen."

GOOD LUCK with your BAD LUCK

Chapter VI
Bad Luck

"I now believe that there's only a certain amount of good luck in the world, and so if something good happens to me, that means something bad has to happen to somebody, somewhere."
— Marshall Brickman, Academy Award-winning screenwriter

As I sat there in anticipation of what he was about to say next, I noticed a slight difference in his facial expression that told me he was a little sad to tell this part of his story. It made me a little sad, too. I waited until he was ready to speak before I said anything. About a minute went by, but this awkward silence felt like an eternity. Then suddenly, he spoke.

"There is a dark side to this story I alluded to earlier about requiring a balance in this equation. You are correct in suggesting that for good luck to occur, there has to be bad luck. The universe must have balance, no matter what. All religions, all cultures, and all scientists believe this to be true in one form or another. It is undisputed. So if you do not believe in good luck, then maybe you believe in bad luck. Whatever you believe in, whatever side of this fence you find yourself on, you cannot dispute the fact that both good luck and bad luck occur at the same time. In my case, that is very true."

"There has been a price for all the championships that my good luck has helped these teams achieve. If we go back to the beginning in 1987, the Twins had won the World Series. That was the first of many championships and the first of many tragedies that I would not come to realize we're connected until much later in my life."

"During that time, a friend of mine was killed in a boating accident, and my grandmother had a heart attack. When West Germany won the World Cup in 1990, my mother's house burned down, and we lost everything. My father died. They sent me to war in Iraq. It was a terrible year for me."

"If we go on to Georgia, I want to bring up the accident I was in where the seat-belt saved my life. My passenger, a buddy from my unit, was not so lucky. He had to have reconstructive surgery on his elbow and hand. We skidded upside down, scraping against the pavement, along with his arm. I also lost my best friend that year in a friendly fire incident. It involved me in dealing with a cover-up by the United States Army. That year was just as bad as the previous year. Look, there were a lot of bad things happening in the places where I worked that affected me or my family members."

"In Florida, the local hospital was committing fraud, and they indicted the local sheriff and his wife for helping Colombians smuggle drugs into the country. I almost killed a man with my bare hands down there. That was a mess. They had a great setup with the smuggling ring. The Colombians would drop the drugs from a plane overhead, and the sheriff would keep the local authorities away from the drop site until the cartel handlers would pick up the drugs. The only reason they got busted and raised red flags: The sheriff's wife had expensive plastic surgery, and they were both spending money like it was as abundant as air. This raised too many questions, so the DEA got involved."

"The medical fraud was also a genius idea until a whistle-blower came along." He told me that back then you could get an itemized list of all medical procedures and supplies that were charged to the patient. "Let's use an orthopedic procedure such as ORIF reduction as an example. The charges would read:"

1. Small Orthopedic set—$2,500

2. Orthopedic pack set—$1,500

3. Ragnall retractors—$250

4. Lap sponges—$125

5. Small Holman retractors—$225

6. Bovie ESU hand piece—$750

    He said that the genius part of the fraud was that they double and tripled billed the same product over and over using different products inside the base price. The Ragnalls and small Holman retractors come in the small orthopedic set, for which you were already getting charged $2,500. The lap sponges and Bovie ESU hand-piece comes in the Orthopedic pack set. Another way to think about it is this: "You order a hamburger from McDonalds. The cash register says it's $1.50, but they end up charging you for the ketchup, bread, mustard, and wrapper separately. I believe that is why they do not give out the itemized list anymore, and over time they have become better at hiding their greedy fraud. After I completed that work assignment, I headed off to Chicago, where the Bulls won their sixth championship in eight years."

    "Something happened in the Chicago area you might remember. A small Cessna plane crashed into the hospital where I was working. I mean, really, a plane crashed into the hospital! Now that was terrible luck for the hospital, but somewhat good luck for the pilot."

"The first couple of weeks I was in New York, they closed down Staten Island because a suspect had shot a cop in the head. I believe the cops also shot the suspect in the head. I worked at a couple of different hospitals in New York City. In another incident at another hospital, a six-year-old boy died because of an oxygen tank that was in the MRI room. As soon as they turned the MRI on, it acted as one gigantic magnet and caused the metal oxygen tank to fly right into the machine and killed the boy. Also, the head of the neurology department at one hospital where I worked tried killing himself by shooting himself in the head."

"So, as you see, bad luck follows good luck, and as far as I can tell, it's a very symbiotic relationship. One cannot live or flourish without the other."

I replied to that by saying to him: "So, you might be the gatekeeper, the balance-maker, the referee of good luck and bad luck."

"I really don't know, but that sounds good."

"My next stop was Texas, where in 2000 and 2001, Enron filed for bankruptcy. So many people lost money, jobs, and their homes. And Tropical Storm Allison in 2001 closed down the Texas hospitals where I worked. Twenty-three people died, and there was five-billion dollars in damage. Also, a hospital in Texas where I worked closed down. Then I went back to Minnesota to go to my grandmother's funeral. Death has followed me and hovered all around me. The other places I worked and lived have fallen victim to the bad-luck side of the equation, too."

"When I was back in New York, the hospital where I worked was about to close down. Their staff had committed seven sentinel events in a six-month period. Sentinel events are things that are never supposed to happen, such as leaving a sponge in a patient's body, cutting off the wrong limb, or prescribing the wrong medication. I want to say the name of the hospital because I think we should warn the public, but I may not. The other hospitals in New York where I worked had high infection rates because of one particular vascular surgeon. His incompetence caused some deaths, but nothing changed because he brought in a lot of money to the hospitals."

"I am surprised when I look back at all the places I have been and worked to realize that there was a correlation between the championships and the bad luck that ensued."

"When I stopped traveling and working to take a break for a while, I lived in the New Jersey/Philadelphia area. I hated that area with a passion. Hated the corruption, hated the attitude. At one point, the entire area could have fallen into the ocean and I wouldn't have cared less."

"Now listen to this next part very carefully. If you don't yet believe in the fifty-eight championships I helped secure because of my good luck, what I am about to tell you should finally convince you. I had gone from being a good luck charm for so many sports franchises to being a curse for an area that I despised, and because of this, no sports team in that area during my stay won a world championship."

"No MLS, MLB, NBA, NFL, NHL, or WNBA championships graced the states of New York or New Jersey or the city of Philadelphia. From 2013 to the beginning of 2016. I also believe there were no college championships for that area as well. Forget about it! It was like a dark cloud hovered over that part of the Eastern Seaboard. They had to wait for the sun to shine back on them. It was a championship blackout. Eventually, the 'sun' reappeared when I left New Jersey and went back to traveling in January 2017. The cloud disappeared. The Philadelphia Eagles ended up winning the 2018 Super Bowl."

This story reminded me of that show, The Twilight Zone. I told him I had to check out this additional information to confirm it. I already knew it would check out because I was a big sports fan. I knew of most of the championship teams, and New York, New Jersey, and Philadelphia were not on that list during the years he lived there. God, this was amazing! I told my friend that this was a great story!

I realize that fifty-eight wins is an impressive number, but I needed to prove it. How could I do that, my friend? He just said to me that he was moving to Oregon and would work there and in Idaho. So, when there is a championship there in the months ahead, then I guess that would be your proof.

"After one assignment, I will take some more time off, so this is a onetime shot for you and your story," he told me.

I said, "Okay, no problem. When do you leave that area?"
"Next January for eight months."
"So, if any team wins in Oregon or Idaho before then, this will prove you are truly a real-life good luck charm."

"I guess so."

"So, I guess we are done."
He agreed and added that we should therefore have some more drinks to celebrate my story. "Hell yes!" I replied.

As the night went on, he filled me in with his wealth of sports knowledge and the answers to many, many sports-trivia questions he had in his overflowing brain. He described his knowledge as useless knowledge, but I was not too sure about that. As I looked at my watch, I realized it was going on 2 a.m. and I'd had way too much to drink to form a coherent sentence, let alone be able to drive home. So I summoned an Uber and said good night to my friend.

I advised him I would call him with any updates about the story's progress and would stay in touch to find out if the next place he went produced a national championship. He laughed like it was no big deal, like it was a sure thing. He was sure about his ability to produce championships. I suppose that after fifty-eight championships comes a little confidence. I was a believer already and hoped that he would take an assignment in Minnesota. Well, I said my goodnights and wished him well on his next adventure. Oh! I almost forgot to mention: My Vikings beat the Chicago Bears that day, so I took it as a good omen.

Chapter VII
Quotes

"Old words are reborn with new faces"
—Criss Jami, Killosophy

While sports are known for producing the most remarkable athletes, colorful characters, influential leaders, and memorable heroes, its fans have only seen a small throng of individuals leave as significant a legacy with their words as with their abilities.

These range from legends to sportswriters and wannabe stars to fiery coaches. These individuals never seemed too hesitate when speaking their minds, even at the expense of grammatical correctness. I picked certain quotes that I thought would interest loyal sports fans who might like to hear what some of the greatest sports minds have had to say about luck. They appear below. But first, I asked my friend to read through some of these quotes I handpicked, and here are some of his responses:

"Nobody paid any attention to Vince Lombardi until he won those five championships. There is a long list of athletes and coaches who we can't remember because they won none. What is that saying? It goes something like this: People always re- member the winners and always forget the runners-up."

"Here's a simple test: Can you name both of the teams who played in the last forty Super Bowls or any other championship games? People remember the winners, and they remember what they have to say."

Here are what past athletes and coaches have had to say — and my friend's comments about their frequently quoted remarks.

"Winning isn't everything, but wanting to win is everything." — Vince Lombardi
"And if you want to win, then just look at the fifty-eight national and world championships I have on my resume, and you decide if you want your team to win." — The M&M Man

"The difference between the impossible and the possible lies in a man's determination." — Tommy Lasorda
"And maybe a little luck on his side." — The M&M Man

"It's just a job. Grass grows, birds fly, waves pound the sand. I beat people up." — Muhammad Ali
"It's what happens. You vote, I go to a country (or state), I work, they win. I bring championships home." — The M&M Man

"Don't measure yourself by what you have accomplished but what you should have accomplished with your ability." — John Wooden
"The fifty-eight championships mean nothing; it's the next one that counts." — The M&M Man

"Show me a guy who's afraid to look bad, and I will show you a guy you can beat every time." — Lou Brock
"Show me a fan, player, or coach who does not believe in good luck, and I will show you a team that has never won a championship." — The M&M Man

"A good coach will make his players see what they can be rather than what they are." — Ara Parseghian
"And luck will make sure you will get there." — The M&M Man

"It's not the will to win that matters—everyone has that. It's the will to prepare to win that matters." — Paul "Bear" Bryant
"That includes luck and believing that it will work in their favor." — The M&M Man

"Persistence can change failure into extraordinary achievement." — Matt Biondi
"Replace persistence with the word luck and you have yourself a world championship." — The M&M Man

"Keep your dreams alive. Understand that to achieve anything requires faith, and belief in yourself, vision, hard work, determination, and dedication. Remember, all things are possible for those who believe." — Gail Devers, two-time Olympic 100-meter champion
"This is good advice. It also applies to believing in good luck. It is a symbiotic relationship." — The M&M Man

"The spirit of sports gives each of us who participates an opportunity to be creative. Sports knows no sex, age, race, or religion. Sports give us all the ability to test our- selves, mentally, physically, and emotionally in a way no other aspect of life can. For many of us who struggle with 'fitting in', sports gives us our first face of confidence. That first bit of confidence can be a gateway to many other great things." — Dan O'Brien
"Therefore sports is the great unifier if used in the right way." — The M&M Man

"Every kid around the world who plays soccer wants to be Pelé. I have a great responsibility to show them not just how to be a soccer player but how to be like a man." — Pelé
"I like this quote the best because it shows what kind of man Pelé is, and it shows he cares about his country and the future of his country." — The M&M Man

"You have to have the mentality of executing your game when you don't feel like there's a lot of hope. I think the best feeling is when somebody pushes you to the limits and you dig down

just a little extra. By the same token, you also need a little luck. Some- times they come together." — Andre Agassi
"That's what I am talking about, motherfuckers." — The M&M Man

"If you can believe it, the mind can achieve it." — Ronnie Lott
"Believe in your good luck charm, man." — The M&M Man

"To succeed... you need to find something to hold on to. Something to motivate you, to inspire you." — Tony Dorsett
"Having a good luck charm, something to rally around, is a plus in the team's column that goes that way." — The M&M Man

"You have to have a lot of luck to win a Super Bowl." — Bill Parcells
"I could never disagree with that statement." — The M&M Man

Chapter VIII
Sports Trivia

I wanted to include this section about sports trivia because as a sports fan enjoy it so much, and I believe that there are many of you who, like me, appreciate trivia such as the trivia I learned from my friend. So, here they are on the next few pages. Just how wide is your knowledge of all things sporting? Find out with these great sports trivia questions and answers. Good luck.

Q. What is the only team to play in every World Cup men's soccer tournament?

A. Brazil

Q. What is soccer player Edson Arantes do Nascimento's better-known nickname?

A. Pelé

Q. Which NFL team appeared in four consecutive Super Bowls from 1991 to 1994 and lost them all?

A. The Buffalo Bills

Q. In 1877, which country's first-ever home football (soccer) international game was played on the grounds of a horse racecourse?

A. Wales

Q. Name the U.S. male volleyball player who won three Olympic gold medals?

A. Karach Kiraly

Q. When was the first baseball game played? 1846 — The New York 9 defeated the

A. 1846 - The New York 9 defeated the Knickerbocker, 23-1

Q. Which is the number between the 5 and the 9 on a British dartboard?

A. 12

Q. Which NFL team appeared in four consecutive Super Bowls and lost them all?

A. The Buffalo Bills

Q. Which team has won the most Super Bowls?

A. tie between the New England Patriots and the Pittsburgh Steelers

Q. Which country won the first-ever World Cup title in 1930?

A. Uruguay

Q. Which country has the most Formula One World Championship winning drivers?

A. The United Kingdom, with ten different winners

Q. Which American football team won the Super Bowls in 1967 and 1968?

A. The Green Bay Packers

Q. At the 1976 Olympic Games in Montreal, gymnast Nadia Comaneci became the first gymnast to score a perfect 10. Which country did she represent?

A. Romania

Q. In which sport did Mark Spitz win seven gold medals at the 1972 Olympics?

A. Swimming (with seven records, one in each race)

Q. In which sport were women first allowed to compete at the Olympics?

A. Tennis

Q. Which was the first African country to qualify for the soccer World Cup?

A. Egypt

Q. On which day of the week do Spanish professional teams usually play soccer?

A. Sunday

Q. What was the nickname of Olympic sprinter Florence Griffith-Joyner?

A. Flo-Jo

Q. Who were the runners-up in the 1966 soccer World Cup?

A. West Germany

Q. In which sport did Ben Hogan win the British Championship?

A. Golf

Q. Which German won the Wimbledon men's singles in 1991?

A. Michael Stich

Q. In which event did Bruce Jenner won the gold medal in the Olympics?

A. The decathlon

Q. What was the age of Boris Becker when he won Wimbledon?

A. 17

Q. What sport uses a black, one-inch thick, three-inch-diameter object that weighs be- tween 5.5 and 6 ounces?

A. Ice hockey (puck)

Q. Who did Muhammad Ali beat to win the World Heavyweight Boxing Championship?

A. Sonny Liston

Q. What color jersey does the leader in the Tour de France wear?

A. Yellow

Q. What is the national sport of Canada?

A. Ice hockey

Q. Has an Arabic country ever won a World Cup?

A. No

Q. What are the dynasty teams in sports (a dynasty is defined as three or more championship wins in a row)?

A. Celtics, Lakers, Bulls, Yankees, Athletics, Canadiens, Packers, Maple Leafs, Islanders, Comets, (New York) Arrows

Q. Which team holds the record for the longest winning streak in college football?

A. The Oklahoma Sooners

Q. What was the largest victory margin in a college football game?

A. 222-0 — Georgia Tech defeated Cumberland College

Q. Who was the last non-quarterback to be named the NFL MVP?

A. Adrian Peterson

Q. Which NFL legend had only one NFL rushing title in his career?

A. Walter Payton

Q. In the modern NFL era, how many days in a week are the games played?

A. Every day of the week

Q. What was the former name of the Chicago Bears?

A. The Chicago Staleys

Q. Which state has produced more pro football Hall of Famers than any other?

A. Pennsylvania

Q. In which city did the LA Rams franchise get started?

A. Cleveland

Q. Who is the only left-handed quarterback in the Pro Football Hall of Fame?

A. Steve Young

Q. Who is known as the king of soccer?

A. Pelé

Q. What is the nickname of the French national soccer team?

A. Les Bleus

Q. Who was the goalkeeper in the film *Victory?* **A. Sylvester Stallone**

Q. Which Mexican city has a soccer team known as the "Goats" (*Chivas*)? **A. Guadalajara**

Q. In which country was the soccer great Ronaldo born?

A. Brazil

Q. How old was Lionel Messi when he played for La Liga?

A. 17

Q. Which player was named BBC African footballer of the year in 2013?

A. Yaya Toure

Q. Anthony Lopes is the goalkeeper for which World Cup squad?

A. Portugal

Q. What is the nickname of the Trinidad and Tobago soccer team?

A. The Soca Warriors

Q. Who was named Danish player of the year four times in his career?

A. Brian Laudrup

Q. Which Spanish club is also known as "The Yellow Submarine?"

A. Villarreal

Q. What team did Pelé play for in the U.S. in the 1970s?

A. The New York Cosmos

Q. Which soccer player is nicknamed the "Czech Cannon?"

A. Pavel Nedved

Q. What year was Pelé's final World Cup appearance?

A. 1970

Q. In 2014, which Nigerian star made the MLS goal of the year?

A. Obafemi Martins

Q. Who was the FIFA World Player of the Year in 2000?

A. Zinedine Zidane

Q. What are the nine legitimate ways to get onto first base in baseball?

1. **Single (any hit)**
2. **Base on balls (walk)**
3. **Fielder's choice**
4. **Hit by pitch**
5. **Fielding error**
6. **Dropped third strike (passed ball or wild pitch)**
7. **Catcher's interference**
8. **Fielder's interference/obstruction**
9. **Batted ball hits another runner before a fielder touches it**

Q. Which college teams do not end in the letter 's'?

Notre Dame Fighting Irish Alabama Crimson Tide

Stanford Cardinal
Illinois Fighting Illini
Syracuse Orange
Navy Midshipmen
North Carolina State Wolfpack Nevada Wolf Pack

Tulane Green Wave Marshall Thundering Herd Tulsa Golden Hurricane North Texas Mean Green

Q. Who was the first player In the NFL to throw a pass, catch it, and score?

A. Brad Johnson (Minnesota Vikings)

Q. In how many countries is soccer played?

A. 200-plus

Q. Which coach came off the bench to play for his team in the Stanley Cup Finals?

A. Lester Patrick

Q. Which NFL player returned a fumble sixty-six yards—to the wrong end zone?

A. Jim Marshall (Minnesota Vikings)

Q. At the 2000 Summer Paralympics, how many members of the gold medal-winning Spanish basketball team were later found to have no disability?

A. 10

Q. Players at which college football position receive the Jim Thorpe Award?

A. Defensive Back

Q. How many knobs, switches, and buttons does a Formula One steering wheel have?

A. 35

Q. Who was the first rookie QB to surpass 3,500 passing yards and throw fewer than five interceptions?

A. Dak Prescott

Q. Which pitcher threw the only no-hit game in World Series history?

A. Don Larsen in the 1956 World Series, Game 5 against the Yankees

Q. Who was the first soccer player to win four European Golden Shoe awards?

A. Cristiano Ronaldo

Q. Which American athlete won four gold medals at the 1936 Summer Olympics in Berlin?

A. Jesse Owens

I hoped you liked some of these. Trivia questions are always good conversation starters, especially if you are a huge sports fan. Remember: Knowledge is power.

Chapter IX
The Follow-up

It has been over two years since I have seen my friend, but we have talked over the phone many times, especially after I found out that Oregon State won the College World Series. God, he was right! He is a legitimate, real-life good luck charm.

I was waiting for a Skype call from my friend to talk about an idea that I had and hoped he would agree to it. After anxiously waiting about ten minutes, I watched some YouTube videos to pass the time. Another five minutes went by, and I was still waiting.

I was about to order a second beer when my iPad rang. It was my mysterious friend finally getting back to me. It looked like he was still in his hospital scrubs. He apologized for being late for the Skype call, but he said they delayed him at work as I figured by his attire. It felt good to see and talk with my friend again.

We exchanged pleasantries and caught up with stories over the two-and-a-half years since I'd seen him. After about thirty minutes of catching up with each other's lives, I could not hide my excitement anymore. I was dying to tell him my good news. He said the draft that I sent to him was pretty well written and thanked me for keeping his name out of it. Well, I thanked him very much for his praise, and then told him he would not like my next proposal. He frowned a little and waited for me to tell him what I was thinking.

I told him I'd received some attention from some sports enthusiast and a publisher about a year earlier. He asked what they wanted. I replied: "They want you. They want me to write a book about you, and I came up with a better Idea. What would you think about a worldwide contest?"

"What are you talking about?" he asked.
I explained that I wanted to write a book about him. I wanted to tell his story and to invite any sports fans who read it to vote for their favorite team by a certain deadline. Whichever team or country got the most votes would be where you will go, I told him, to take an assignment, live, work, and do what you do to bring home a championship for the home team.

He laughed and replied, "Are you fucking serious?"

"I am absolutely serious about this. You have a gift; you have something that nobody can explain in any rational way. It works. Your good luck just seems to work, and I want to give the world a chance to see it."

"Think of all the World Cup teams that have never won a World Championship or teams like the Cleveland Browns, Detroit Lions, and my beloved Minnesota Vikings. The publisher will pay for all your travel and housing. Nobody will know who you are until you complete your assignment, and if we can keep this a secret, too, we will try. This can be one of the biggest stories in the history of sports. When you bring a world championship to some team that needs it, deserves it, or wants it, you will have completed your journey."

"I have a question for you: Is this more important to you, the owner, or the fan? I replied by saying it is for the fan. The fan does not make any money, get any glory, or get any recognition when their teams win. So don't do this for the owners and don't do it for the players. Don't do it for me or the publisher. Do it for the millions of fans all over the world. Put the power in the hands of the fans and let them vote and decide where you go and who will win that next world title."

There was his characteristic awkward silence again, this time lasting for five long minutes. He finally replied with just one question: "Why do you want me to do this?"

I sat there in silence for a lengthy time. My thoughts went everywhere. I was wondering what value there was in all of this. Who determines value, what determines value? I thought about all of this for a long while, and I came up with some pretty interesting answers. First, the dictionary definition of value is:

1.) It holds the regard that something to deserve; the importance, worth, or use fulness of something.

2.) Consider (someone or something) to be important or beneficial; have a high opinion of.

The dictionary refers to desirability, beauty, accessibility, age, condition, and sentiment. So, with all of this information, does this mysterious man have any value? According to the dictionary and man's definition, the answer is an astounding yes.

My friend has value because of the importance of his accomplishments. Fifty-eight championships, whether it was a miracle, luck, or something else, are of importance because of the sheer number of championships he has helped produce. Then the question becomes, is he worth something? Of course he is. Casting a vote for your favorite team in the hopes of a championship for your favorite team, I would say, is worth a hell of a lot. How many teams have picked high in the draft and paid multi-millions of dollars for players that in the end continuously disappoint us. My Vikings are a prime example. The Herschel Walker trade was the worst sports deal in the history of sports. That deal propelled the Dallas Cowboys to three Super Bowls in the Nineties. In some respects, you could say the Minnesota Vikings have won three Super Bowls through osmosis. Another Vikings blunder is the $84 million quarterback deal. What has it gotten us? Nothing but pain and suffering. There are other teams that fit this bill. The Detroit Lions, Cleveland Browns, and Buffalo Bills to name just a few. So, the worth of having an authentic good luck charm on your team costs nothing and is worth everything.

Now we come to the question: Is he useful? Well, that is a straightforward answer as well. He is useful because his luck will drum up excitement for all major sport teams around the world like we have never seen before. It will give players a psychological advantage in the playoffs to know the fans went out there and voted for them to win. We know athletes play better when they have a good luck charm, or an outside influence guiding them, or inspiring them.

Is my friend a rarity? Fuck yes! I know of no other good luck streak like his. I have researched and researched and found nothing. Imagine this for a second: Let's say he goes to a state such as Michigan, New York, or Minnesota and then the Detroit Lions, Buffalo Bills, or Minnesota Vikings win a Super Bowl. How important, rare, and desirable will he be then?

So now the final question is: Would he be valued as a pseudo member of any sports team that wants to win a national and/or world championship? The simple and only answer is yes. The question that has to be answered now is: What team is deserving of this real-life good luck charm? Only the fans can decide that. I came out of my comatose state to answer him.

"I am hoping there are a lot of Vikings fans out in the world," I began. "I believe in you! I like underdogs winning, and if you can help a team like Mexico, Peru, or even Qatar win a World Cup for the first time, or like I said, bring a title to the Browns, Lions, or even my Vikings, then hell yes I want to see it! I want to live it! I want to experience it, and I know I am not the only one."

"Everyone needs a wonderful story; they need faith; they need to believe. And when they do, and that belief and faith are rewarded with a world championship, there is nothing else like it in the sports world for a fan." There was another awkward silence, but this time I was the one to speak after the long silence, and all I said was, "What say you, my friend?

Chapter X

A List of Major Sports Leagues & Teams (including NCAA Division I schools)

I compiled a list of all the major sports leagues and which teams have won, come close, or never had the chance to win a national or world championship. There are a lot of teams on this list that deserve a better chance of winning that elusive crown than they have been given. The first list is of Division I college teams categorized by state:

ALABAMA

University of Alabama	28 Championships
Alabama A&M University	00 Championships
Alabama State University	00 Championships
Alabama Agricultural & Mechanical University	05 Championships*
University of Alabama at Birmingham	00 Championships
Auburn University	27 Championships*
Jacksonville State University	00 Championships
Samford University	00 Championships
University of North Alabama	07 Championships*
University of South Alabama	00 Championships
Troy University	16 Championships*

ALABAMA

Tuskegee University	15 Championships*

ARIZONA

University of Arizona	19 Championships*
Arizona State	46 Championships*
Northern Arizona University	03 Championships*
Grand Canyon University	02 Championships*

ARKANSAS

Arkansas State University	00 Championships*
University of Arkansas at Little rock	00 Championships
University of Arkansas at Pine Bluff	01 Championships*
University of Central Arkansas	00 Championships
University of Arkansas	48 Championships

CALIFORNIA

California Baptist University	00 Championships
California Polytechnic State University	36 Championships*
California State University at Bakersfield	30 Championships*
California State University at Frenso	04 Championships*
California State University at Fullerton	13 Championships*
California State University at Long Beach	15 Championships*
California State University at Sacramento	04 Championships*
University of California at Berkeley	43 Championships*

CALIFORNIA

University of California at Davis	08 Championships*
University of California at Irvine	22 Championships*
University of California at Los Angeles	129 Championships*
University of California at Riverside	05 Championships*
University of California at Santa Barbara	02 Championships*
University of California at San Diego	03 Championships*
Long Beach State University	03 Championships*
Loyola Marymount University	00 Championships
University of the Pacific	00 Championships
Pepperdine University	13 Championships*
Saint Mary's College of California	00 Championships
University of San Diego	00 Championships
San Diego State University	15 Championships*
University of San Francisco	00 Championships
San Jose State University	00 Championships
Santa Clara University	00 Championships*
University of Southern California	00 Championships*
Stanford University	163 Championships

COLORADO

University of Colorado at Boulder	29 Championships*
Colorado State University	00 Championships

COLORADO

University of Denver	37 Championships*
University of Northern Colorado	00 Championships
U.S. Air Force Academy	02 Championships*

CONNECTICUT

Central Connecticut State University	00 Championships
University of Connecticut	23 Championships*
University of Hartford	00 Championships
Fairfield University	00 Championships
Quinnipiac University	00 Championships
Sacred Heart University	00 Championships*
Yale University	56 Championships*

DELAWARE

University of Delaware	17 Championships*
Delaware State University	01 Championships*

D.C.

American University	00 Championships
George Washington University	00 Championships
Georgetown University	03 Championships
Howard University	09 Championships*

FLORIDA

Bethune-Cookman University	00 Championships
University of Central Florida	02 Championships*
University of Florida	41 Championships*

FLORIDA

Florida A&M University	15 Championships
Florida Atlantic University	00 Championships
Florida Gulf Coast university	00 Championships
Florida International University	00 Championships
Florida State University	19 Championships*
Jacksonville University	00 Championships
University of Miami	21 Championships*
University of North Florida	02 Championships*
University of South Florida	09 Championships*
Stetson University	00 Championships

GEORGIA

University of Georgia	45 Championships*
Georgia Institute of Technology	05 Championships*
Georgia Southern University	06 Championships*
Georgia State University	00 Championships
Kennesaw state University	05 Championships*
Mercer University	00 Championships*
Savannah State University	00 Championships*

HAWAII

University of Hawaii at Manoa	04 Championships*

IDAHO

Boise State University	04 Championships*
University of Idaho	00 Championships

IDAHO

Idaho State University	00 Championships

ILLINOIS

Bradley University	00 Championships
Chicago State University	00 Championships
DePaul University	00 Championships
Eastern Illinois University	06 Championships*
Illinois State University	01 Championships*
University of Illinois at Urbana-Champaign	23 Championships*
University of Illinois at chicago	00 Championships
Loyola University Chicago	00 Championships
Northern Illinois University	00 Championships
Northwestern University	13 Championships*
Southern Illinois University at Carbondale	01 Championships*
Southern Illinois University at Edwardsville	17 Championships*
Western Illinois University	00 Championships

INDIANA

ball State University	00 Championships
Butler University	00 Championships
University of Evansville	00 Championships
Indiana State University	03 Championships*
Indiana University at Bloomington	26 Championships*
Indiana University-Purdue University at Fort Wayne	00 Championships

INDIANA

Indiana University-Purdue University at Indianapolis	00 Championships
University of Notre Dame	20 Championships*
Purdue University	05 Championships*
Valparaiso University	00 Championships

IOWA

Drake University	06 Championships
University of Iowa	32 Championships*
Iowa State University	20 Championships
University of Northern Iowa	00 Championships

KANSAS

University of Kansas	11 Championships*
Kansas State University	00 Championships
Wichita State University	00 Championships

KENTUCKY

Bellarmine University	00 Championships
Eastern Kentucky University	00 Championships
University of Kentucky	11 Championships
More Head State University	00 Championships
Murray State University	02 Championships*
Northern Kentucky University	03 Championships*
Western Kentucky University	00 Championships

LOUISIANA

University of Louisiana at Lafayette	00 Championships

LOUISIANA

University of Louisiana at Monroe	01 Championships
Louisiana State University	48 Championships*
Louisiana Tech University	06 Championships
Grambling State University	16 Championships*
McNeese State University	00 Championships
University of New Orleans	00 Championships
Nicholls State University	00 Championships
Northwestern state University	00 Championships
Southeastern Louisiana University	00 Championships
Southern University at Baton Rouge	11 Championships*
Tulane University	02 Championships*

MAINE

University of Maine	02 Championships

MARYLAND

	17 Championships*
Coppin State University	00 Championships
Johns Hopkins (lacrosse)	54 Championships*
Loyola University Maryland	00 Championships
University of Maryland Eastern Shore	00 Championships
University of Maryland, Baltimore County	00 Championships
University of Maryland, College Park	47 Championships
Morgan State University	00 Championships
Mount St. Mary's University	00 Championships

MARYLAND

Towson University	00 Championships
U.S. Naval Academy	73 Championships*

MASSACHUSETTS

Boston College	05 Championships*
Boston University	04 Championships
Harvard University	13 Championships*
College of the Holy Cross	04 Championships*
University of Massachusetts at Amherst	00 Championships
University of Massachusetts at Lowell	07 Championships*
Merrimack College	04 Championships
Northeastern University	00 Championships

MICHIGAN

Central Michigan University	00 Championships
University of Detroit Mercy	00 Championships
Eastern Michigan University	16 Championships*
University of Michigan	58 Championships
Michigan State University	30 Championships
Oakland University	10 Championships*
Western Michigan University	02 Championships

Minnesota

University of Minnesota, Twin Cities	41 Championships*

MISSISSIPPI

Alcorn State University	05 Championships*

MISSISSIPPI

Jackson State University	03 Championships*
University of Mississippi	03 Championships
Mississippi State University	00 Championships
Mississippi Valley State U	00 Championships
The University of S.Mississippi	00 Championships

MISSOURI

Missouri State University	03 Championships*
University of Missouri at Columbia	02 Championships
University of Missouri at Kansas City	02 Championships
Southeast Missouri State University	00 Championships
Saint Louis University	00 Championships

MONTANA

University of Montana at Missoula	02 Championships*
Montana State University at Bozeman	00 Championships

NEBRASKA

Creighton University	01 Championships
University of Nebraska, Omaha	11 Championships
University of Nebraska, Lincoln	28 Championships

NEVADA

University of Nevada at Las Vegas	02 Championships
University of Nevada at Reno	00 Championships

NEW HAMPSHIRE

Dartmouth College	21 Championships*
University of New Hampshire	00 Championships

NEW JERSEY

Fairleigh Dickinson University at Metropolitan Campus	00 Championships*
Monmouth University	00 Championships
New Jersey Institute of Technology	00 Championships
Princeton University	24 Championships
Rider University	00 Championships
Rutgers, The State University of N.J. at New Brunswick	02 Championships
Seton Hall University	00 Championships
Saint Peter's University	00 Championships

NEW MEXICO

University of New Mexico	03 Championships
New Mexico State University	00 Championships

NEW YORK

University at Albany	00 Championships
Binghamton University	00 Championships
University at Buffalo, the State University of New York	00 Championships
Canisius College	00 Championships
Colgate University	00 Championships
Columbia University	38 Championships
Cornell University	78 Championships

NEW YORK

Fordham University	00 Championships
Hofstra University	00 Championships
Iona College	00 Championships
Long Island University-Brooklyn Campus	00 Championships*
Manhattan College	00 Championships
Marist College	00 Championships
Niagara University	00 Championships
Siena College	00 Championships
St. Bonaventure University	00 Championships
St. Francis College Brooklyn	00 Championships
St. John's University(New York)	02 Championships
Stony Brook University	00 Championships
Syracuse University	15 Championships
West Point	32 Championships
Wagner College	00 Championships

NORTH CAROLINA

Appalachian State University	03 Championships*
Campbell University	00 Championships
Davidson College	00 Championships
Duke University	17 Championships
East Carolina University	00 Championships
Elon University	00 Championships
Gardner-Webb University	00 Championships

NORTH CAROLINA

High Point University	00 Championships
University of North Carolina-Ashville	00 Championships
North Carolina A&T State University	00 Championships
North Carolina Central University	00 Championships*
North Carolina State University	00 Championships
North Carolina Agricultural & Technical State University	01 Championships
University of North Carolina-Wilmington	00 Championships
University of North Carolina-Chapel Hill	45 Championships
University of North Carolina-Charlotte	00 Championships
University of North CarolinaGreensboro	05 Championships*
Wake Forest	09 Championships
Western Carolina University	00 Championships

NORTH DAKOTA

University of North Dakota	08 Championships
North Dakota State University	00 Championships

OHIO

University of Akron	00 Championships
Bowling Green State University	01 Championships
University of Cincinnati	06 Championships
Cleveland State University	00 Championships
University of Dayton	00 Championships
Kent State University	00 Championships
Miami University (Ohio)	00 Championships

NORTH CAROLINA

The Ohio State University	79 Championships*
Ohio University	00 Championships
University of Toledo	00 Championships
Wright State University	00 Championships*
Xavier University	01 Championships
Youngstown State University	04 Championships

OKLAHOMA

University of Oklahoma	38 Championships
Oklahoma State University	53 Championships
Oral Roberts University	00 Championships
The University of Tulsa	04 Championships*

OREGON

University of Oregon	33 Championships
Oregon State University	04 Championships
University of Portland	03 Championships
Portland State University	07 Championships

PENNSYLVANIA

Bucknell University	01 Championships
Drexel University	02 Championships
Duquesne University	00 Championships
La Salle University	02 Championships
Lafayette College	00 Championships
Lehigh University	00 Championships
University of Pennsylvania	53 Championships

PENNSYLVANIA

Pennsylvania State University	97 Championships
University of Pittsburgh	11 Championships
Robert Morris University	00 Championships
Saint Francis University (Pennsylvania)	00 Championships
Saint Joseph's University	00 Championships
Temple University	04 Championships*
Villanova	21 Championships

RHODE ISLAND

Brown University	07 Championships*
Bryant University	00 Championships
Providence College	03 Championships
University of Rhode Island	01 Championships

SOUTH CAROLINA

College of Charleston(South Carolina)	00 Championships
Charleston Southern University	00 Championships
The Citadel	00 Championships
Clemson University	06 Championships
Coastal Carolina University	00 Championships
Furman University	00 Championships
Presbyterian College	00 Championships
University of South Carolina, Columbia	04 Championships
South Carolina State University	02 Championships
University of South Carolina upstate	00 Championships*

SOUTH CAROLINA

Winthrop University	00 Championships
Wofford	00 Championships

SOUTH DAKOTA

University of South Dakota	00 Championships
South Dakota State University	08 Championships*

TENNESSEE

Austin Peay State University	00 Championships
Belmont University	00 Championships
East Tennessee State University	00 Championships
Lipscomb University	00 Championships
University of Memphis	00 Championships
Middle Tennessee State University	02 Championships
Tennessee state University	24 Championships*
Tennessee Technological University	00 Championships
University of Tennessee at Chattanooga	23 Championships
University of Tennessee, Knoxville	21 Championships
University of Tennessee at Martin	02 Championships
Vanderbilt University	05 Championships

TEXAS

Abilene Christian University	00 Championships
Baylor University	04 Championships
University of Houston	17 Championships
Houston Baptist University	00 Championships

University of the Incarnate Word	00 Championships
Lamar University	02 Championships
University of North Texas	00 Championships
Prairie View A&M University	00 Championships
Rice University	00 Championships
Sam Houston State University	00 Championships
Southern Methodist University	04 Championships
Stephen F. Austin State University	03 Championships
Tarleton State University	00 Championships
Texas A&M University, College Station	17 Championships
Texas A&M University, Corpus Christi	00 Championships
Texas Christian University	04 Championships
University of Texas Rio Grande Valley	00 Championships
Texas Southern University	01 Championships
Texas State University	00 Championships
Texas Tech University	02 Championships
University of Texas at Arlington	00 Championships
University of Texas at Austin	56 Championships
University of Texas at El Paso	21 Championships
University of Texas at San Antonio	00 Championships

UTAH

Brigham Young University	11 Championships
Dixie State University	00 Championships
Southern Utah University	00 Championships

UTAH

University of Utah	26 Championships
Utah State University	03 Championships
Utah Valley University	00 Championships
Weber State University	00 Championships

VERMONT

University of Vermont	00 Championships

VIRGINIA

George Mason University	02 Championships
Hampton University	01 Championships
James Madison University	05 Championships
Liberty University	09 Championships
Longwood University	00 Championships
Norfolk State University	00 Championships
Old dominion University	00 Championships
Radford University	00 Championships
University of Richmond	00 Championships
University of Virginia	28 Championships
Virginia Commonwealth University	00 Championships
Virginia Military Institute	00 Championships
Virginia Polytechnic Institute & State University	00 Championships
College of William and Mary	02 Championships

WASHINGTON

Eastern Washington University	00 Championships

WASHINGTON

Gonzaga University	00 Championships
Seattle University	00 Championships
University of Washington	51 Championships
Washington State	02 Championships

WEST VIRGINIA

Marshall University	02 Championships
West Virginia University	25 Championships

WISCONSIN

Marquette University	01 Championships
University of Wisconsin, Green Bay	00 Championships
University of Wisconsin, Madison	51 Championships
University of Wisconsin, Milwaukee	29 Championships

WYOMING

University of Wyoming	04 Championships

    If there are any championships missing, it's because they did not meet the publishing deadline or did not meet the criteria for the 58 previous championships so far. I would like to give anybody a chance to correct or add a championship we might have missed. When you vote, leave a comment on our website and we will make the necessary changes if it meets our criteria.

*There are teams that have won championships in Division I and Division II. And teams that have won championships that are not bestowed on them by the NCAA.

*There are also all-black colleges that have won championships. And Ivy League school (CIAA) champions.

* Also, some schools are listed as Division I schools but also participate in some Division II or III sports that predate the NCAA.

National Football League:

TEAM	Championship Games Won/Lost
Arizona Cardinals	0 - 2
Atlanta Falcons	0 - 1
Baltimore Ravens	2 - 0
Buffalo Bills	0 - 4 in a row
Carolina Panthers	0 - 2
Chicago Bears	1 - 1
Cincinnati Bengals	0 - 2
Cleveland Browns	Never
Dallas Cowboys	5 - 3
Denver Broncos	3 - 5
Detroit Lions	Never
Green Bay Packers	4 - 1
Houston Texans	Never
Indianapolis/Baltimore Colts	2 - 2
Jacksonville Jaguars	Never
Kansas City Chiefs	2 - 1
San Diego/LA Chargers	0 - 1
St. Louis/Los Angeles Rams	1 - 3
Miami Dolphins	2 - 3

TEAM	Championship Games Won/Lost
Minnesota Vikings	0 - 4
New England Patriots	6 - 5
New Orleans Saints	1 - 0
New York Giants	4 - 1
New York Jets	1 - 0
Oakland Raiders	3 - 2
Philadelphia Eagles	1 - 2
Pittsburgh Steelers	6 - 2
San Francisco 49ers	1 - 2
Seattle Seahawks	1 - 0
Tampa Bay Buccaneers	2 - 0
Tennessee Titans	0 - 1
Washington Redskins	3 - 2

There is a strange coincidence inside the NFL that involves the Minnesota Vikings. Has anyone ever heard of the Minnesota Vikings being the elephant graveyard for aging quarterbacks? Instead of aging quarterbacks going to the Vikings to die, they revive their career and have their best or close to their best statistical years in their careers. Let's start with the great Fran Tarkenton. He was a Viking beginning in 1961, traded to the NY Giants in 1967, and traded back to the Vikings in 1972, when he had his best years of his career, going to the Super Bowl three times. He was 32 when he began his second stint with the Vikings and retired at age 38. The next elephant quarterback to grace the fields of the Minnesota tundra was Joe Kapp, the first of two 32-year-olds to bring the Vikings to the Super Bowl, and only then falling short of the national title. Joe Kapp had his best year in 1969 and retired in 1971. We now head to Gary Cuozzo, who had his best years with the Minnesota Vikings in 1970 and 1971 before he retired in 1973 at the young age of 32.

After Fran Tarkenton left the Vikings in 1978, it all went downhill until a flash in the pan named Wade Wilson, and Rich Gannon each had a couple good years, and then we had a bunch of older quarterbacks come through Minnesota: Warren Moon becoming the first African-American quarterback and the first un-drafted quarterback. He is also the only player inducted to both the Pro Football Hall of Fame and the Canadian Football Hall of fame to receive the honor. Randall Cunningham enjoyed the strongest season of his career and helped the team set the NFL record for most points in a regular season at the time, although the Vikings would be upset in the the NFC title game. Jeff George, Gus Frerotte, and finally Brett Favre. All of whom had exceptional years with the Vikings and broke NFL records with them. I am just waiting for Tom Brady or Russell Wilson to come through Minnesota.

Major League Baseball:

TEAM	Championship Games Won/Lost
Atlanta Braves	3-6
Arizona Diamondbacks	1-0
Baltimore Orioles	3-4
Boston Red Sox	9-4
Chicago Cubs	3-8
Chicago White Sox	3-2
Cincinnati Reds	5-4
Cleveland Indians	2-4
Colorado Rockies	0-1
Detroit Tigers	4-7
Houston Astros	1-2
Kansas City Royals	2-2
Los Angeles Angels	1-0
LA Dodgers	6-14
Miami Marlins	2-0
Milwaukee Brewers	0-1
Minnesota Twins(1)	3-3
NY Mets	2-3
NY Yankees	27-13
Oakland Athletics	9-5
Philadelphia Phillies	2-5

TEAM	Championship Games Won/Lost
Pittsburgh Pirates	5-2
San Diego Padres	0-2
San Francisco Giants	8-12
Seattle Mariners	Never
St. Louis Cardinals	11-8
Tampa Bay Rays	0-1
Texas Rangers	0-2
Toronto Blue Jays	2-0
Washington Nationals	1-0

(1) 1901-1960 Washington Senators / 1961-present Minnesota Twins
(2-12) LA Dodgers
(2) Brooklyn Atlantics(1884)
(3) Brooklyn Grays(1885-1887)
(4) Brooklyn Bridegrooms(1888-1890)
(5) Brooklyn Grooms(1891-1895)
(6) BrooklynBridegrooms(1896-1898)
(7) Brooklyn Superbas(1899-1910)
(8) Brooklyn Trolley Dodgers(1911-1912)
(9) Brooklyn Dodgers(1913)
(10) Brooklyn Robins(1914-1931)
(11) Brooklyn Dodgers(1932-1957)
(12) Los Angeles Dodgers(1958-Present)

Major League Soccer:

TEAM	Championship Games Won/Lost
Atlanta United FC	1-0
Chicago Fire	1-2
Colorado Rapids	1-1
Columbus Crew SC	1-1
DC United	4-1
FC Dallas	1-0
Houston Dynamo	2-2
LA Galaxy	5-4
Los Angeles FC	Never
Minnesota United	Never
Montreal Impact	Never
New England Revolution	0-5
New York City FC	Never
New York Red Bulls	0-1
Orlando City SC	Never
Philadelphia Union	Never
Portland Timbers	1-1
Real Salt Lake	1-1
San Jose Earthquakes	2-0
Seattle Sounders FC	2-1

TEAM	Championship Games Won/Lost
Sporting Kansas City	2-1
Toronto FC	1-2
Vancouver Whitecaps FC	Never

National Basketball Association:

TEAM	Championship Games Won/Lost
Atlanta Hawks	1-3
Boston Celtics	17-4
Brooklyn Nets(1)	0-2
Charlotte Hornets	Never
Chicago Bulls	6-0
Cleveland Cavaliers	1-4
Dallas Mavericks	1-1
Denver Nuggets	Never
Detroit Pistons(2)	3-4
Golden State Warriors(3)	6-5
Houston Rockets	2-2
Indiana Pacers	0-1
LA Clippers	Never
LA Lakers(4)	16-15
Memphis Grizzlies	Never
Miami Heat	3-2
Milwaukee Bucks	1-1
Minnesota Timberwolves	Never
New Orleans Pelicans	Never
New York Knicks	2-6
Oklahoma City Thunder(5)	1-3

TEAM	Championship Games Won/Lost
Orlando Magic	0-2
Philadelphia Sixers(6)	3-6
Phoenix Suns	0-2
Portland Trail Blazers	1-2
Sacramento Kings(7)	1-0
San Antonio Spurs	5-1
Toronto Raptors	1-0
Utah Jazz	0-2
Washington Wizards(8)	1-3

- (1) Includes record as New Jersey Nets
- (2) Includes record as Fort Wayne Pistons
- (3) Includes record as Philadelphia & San Francisco Warriors
- (4) Includes record as Minneapolis Lakers
- (5) Includes record as Baltimore & Washington Bullets
- (6) Includes record as Syracuse Nationals
- (7) Includes record as Rochester Royals
- (8) Includes record as Baltimore & Washington Bullets

National Hockey League:

TEAM	Championship Games Won/Lost
Anaheim Ducks	1-1
Arizona Coyotes	Never
Boston Bruins	6-14
Buffalo Sabres	0-2
Calgary Flames	1-2
Carolina Hurricanes	1-1
Chicago Blackhawks	6-7
Colorado Avalanche	2-0
Columbus Blue Jackets	Never
Dallas Stars	1-3
Detroit Redwings	11-13
Edmonton Oilers	5-2
Florida Panthers	0-1
LA Kings	2-1
Minnesota Wild	Never
Montreal Canadiens	24-9
Nashville Predators	0-1
New Jersey Devils	3-2
NY Islanders	4-1
NY Rangers	4-7
Ottawa Senators	0-1

TEAM	Championship Games Won/Lost
Philadelphia Flyers	2-6
Pittsburgh Penguins	5-1
San Jose Sharks	0-1
St. Louis Blues	1-3
Tampa Bay Lighting	1-1
Toronto Maple Leafs	13-8
Vancouver Canucks	0-3
Vegas Golden Knights	0-1
Washington Capitals	1-1
Winnipeg Jets	Never

Woman's National Basketball Association:

TEAM	Championship Games Won/Lost
Atlanta Dream	0-3
Charlotte Sting(1)	0-1
Chicago Sky	0-1
Connecticut Sun	0-3
Houston Comets(2)	4-0
Indiana Fever	1-2
Los Angeles Sparks	3-2
Minnesota Lynx	4-2
New York Liberty	0-4
Phoenix Mercury	3-1
Sacramento Monarchs(3)	1-1
San Antonio Silver Stars(4)	0-1
Seattle Storm	3-0
Tulsa Shock(5)	3-1
Washington Mystics	1-1

- (1) Folded following 2006 season
- (2) Folded following 2008 season
- (3) Folded following 2009 season
- (4) Relocated to Las Vegas following 2017 season
- (5) Relocated to Tulsa following 2009 season and Dallas following 2015 season

The Federation international Football Association:

TEAM (Country)	Championship Games Won/Lost	Never Been in a Championship Game
Afghanistan		0
Albania		0
algeria		0
American Samoa		0
Andorra		0
Angola		0
Anguilla		0
Antigua & Barbuda		0
Argentina	2-3	
Armenia		0
Aruba		0
Australia		0
Austria		0
Azerbaijan		0
Bahamas		0
Bahrain		0
Bangladesh		0
Barbados		0
Belarus		0

TEAM (Country)	Championship Games Won/Lost	Never Been in a Championship Game
Belgium		0
Belize		0
Benin(1986-)Dahomey		0
Bermuda		0
Bhutan		0
Bolivia		0
Bosnia & Herzegovina		0
Botswana		0
Brazil	5-2	
British Virgin Islands		0
Brunei		0
Bulgaria		0
BurkinaFaso(1990) Upper Volta(1978)		0
Burundi		0
Cambodia		0
Cameroon		0
Canada		0
Cape Verde		0
Cayman Islands		0

TEAM (Country)	Championship Games Won/Lost	Never Been in a Championship Game
Central African Republic		0
Chad		0
Chile		0
China PR		0
Chinese Taipei(1982-) Republic of China(1978)		0
Columbia		0
Comoros		0
Congo		0
Cook Islands		0
Costa Rica		0
Croatia	0-1	
Cuba		0
Curacao(2014-) NetherlandsAntilles (1962-2010) Territory of Curacao(1934-1994)		0
Cyprus		0
Czech Republic(1998-) Representation of Czechs & Slovaks(1994) Czechoslovakia (1934-1994)	0-2	

TEAM (Country)	Championship Games Won/Lost	Never Been in a Championship Game
Denmark		0
djibouti		0
Dominica		0
Dominican Republic		0
DR Congo(1998-) Zaire(1974-1998)		0
East Germany		0
Equador		0
Egypt		0
El Salvador		0
England	1-0	0
Equatorial Guinea		0
Eritrea		0
Estonia		0
Eswatini(2022-) Swaziland(1994-2018)		0
Ethiopia		0
Faroe Islands		0
Fiji		0
Finland		0
France	2-1	0

TEAM (Country)	Championship Games Won/Lost	Never Been in a Championship Game
Gabon		0
Gambia		0
Georgia		0
Germany(1994-) WestGermany(1954-1990 Germany(1934-1938)	4-4	
Ghana		0
Gibraltar		0
Greece		0
Grenada		0
Guam		0
Guatemala		0
Guinea		0
Guinea-Bissau		0
Guyana		0
Haiti		0
Honduras		0
Hong Kong		0
Hungary	0-2	
Iceland		0
India		0

TEAM (Country)	Championship Games Won/Lost	Never Been in a Championship Game
Indonesia(1958-) Dutch East Indies(1938)		0
Iran		0
Iraq		0
Israel(1950-) Palenstine,British Mandate(1934-1938		0
Italy	4-2	
Ivory Coast		0
Jamaica		0
Japan		0
Jordan		0
Kazakhstan		0
Kenya		0
kosovo		0
Kuwait		0
Kyrgyzstan		0
Laos		0
Latvia		0
Lebanon	0-2	
Lesotho		0
Liberia		0

TEAM (Country)	Championship Games Won/Lost	Never Been in a Championship Game
Libya		0
Liechtenstein		0
Lithuania		0
Luxembourg		0
Macau		0
Madagascar		0
Malawi		0
Malaysia		0
Maldives		0
Mali		0
Malta		0
Mauritania		0
Mauritius		0
Mexico		0
Moldova		0
Mongolia		0
Montenegro		0
Montserrat		0
Morocco		0
Mozambique		0

TEAM (Country)	Championship Games Won/Lost	Never Been in a Championship Game
Myanmar		0
Nambia		0
Nepal		0
Netherlands	0-3	
New Caledonia		0
New Zealand		0
Nicaragua		0
Niger		0
Nigeria		0
North Korea		0
North Macedonia(2022-) Macedonia(1998-2018)		0
Northern Ireland(1954-) Ireland(1950)		0
Norway		0
Oman		0
Pakistan		0
Palestine		0
Panama		0
Papua New Guinea		0
Paraguay		0
Peru		0

TEAM (Country)	Championship Games Won/Lost	Never Been in a Championship Game
Philipines		0
Poland		0
Portugal		0
Puerto Rico		0
Qatar		0
Republic of Ireland(1954-) Ireland(1950) Irish free states (1934-1938)		0
Romania		0
Russia(1994-) Soviet Union(1958-1990)		0
Rwanda		0
Saar		0
Saint Vincent & the Grenadines		0
Samoa(2002-) Western Samoa(1998)		0
San Marino		0
Sao Tome & Principe		0
Saudi Arabia		0
Scotland		0
Senegal		0

TEAM (Country)	Championship Games Won/Lost	Never Been in a Championship Game
Serbia(2010-) Serbia & Montenegro(2006) FR Yugoslavia(1998-2002) SFR Yugoslavia(1950-1990) Kingdom of Yugoslavia(1934-1938)		0
Seychelles		0
Sierra Leone		0
Singapore		0
Slovakia		0
Slovenia		0
Solomon Islands		0
Somalia		0
South Africa		0
South Korea		0
South Sudan		0
South Yemen		0
Spain	1-0	
Sri Lanka		0
Sudan		0
Suriname(1978-) Dutch Guyana(1962-1974)		0

TEAM (Country)	Championship Games Won/Lost	Never Been in a Championship Game
Swedan	0-1	
Switzerland		0
Syria		0
Tahiti		0
Tajikistan		0
Tanzania		0
Thailand		0
Timor-Leste		0
Togo		0
Tongo		0
Trinidad & Tobago		0
Tunisia		0
Turkey		0
Turkmenistan		0
Turks & Caicos Islands		0
Tuvalu		0
U.S. Virgin Islands		0
Uganda		0
Ukraine		0
United Arab Emirates		0
United States		0

TEAM (Country)	Championship Games Won/Lost	Never Been in a Championship Game
Uruguay	2-0	
Uzbekistan		0
Vanuatu		0
Venezuela		0
Vietnam(1994-) South Vietnam(1974)		0
Wales		0
Yemen(1994-) North Yemen(1986-1990)		0
Zambia		0
Zimbabwe(1982-) Rhodesia(1970)		0

The amazing thing about the FIFA list is that only eight teams out of 211 have ever won the title of world champion. I believe it is time to change that statistic.

As you scrolled through these lists and saw your team or your alma mater, you may have noticed that your team might of never won a championship. Regarding this launch of the contest I proposed to my lucky friend, maybe you thought, This is my chance to do something about it. Some people and some teams are just born with luck. Everything they touch turns to gold, while it burdened others with some bad luck. When the contest is launched, that will be your time to act! This opportunity will not come around again.

Consider the Carrington Event, a geomagnetic storm in 1859 caused by a massive solar flare. Consider the 1814 flood of beer when one of the iron rings holding a fermentation tank snapped, causing it to burst, releasing over 1.2 million liters

(320,000 gallons) of fermenting beer. A beer lover ruled that an act of God (probably). Consider a dart thrown at a dartboard in a pub that kills a mosquito. These events occur once in a lifetime, if at all. Very few people get to witness a once-in-a-lifetime event. But you have a chance to not only witness but take part in such an event. Joe Namath assured the New York City press the Jets were going to win the Super Bowl, and then they did. What did people think of this fortune teller-football player? People make up their own minds what to believe and what not to believe. Well, this once-in-a-lifetime event is, by definition, extremely rare, especially as an opportunity for the fan to take part.

To paraphrase what a wiser man than I am (my lucky friend) told me: "Hard work will get you to the next level or a juicy fat contract. You need both work and luck for a top draft pick, a first-team spot, or a private jet." His point was that certain accomplishments are within the reasonable grasp of someone who makes incremental gains each day. Outsized success and outlier accomplishments require that in addition to extreme luck or timing. This is worth considering for all of us who grew up being told the world was a meritocracy. Of course, it isn't. Plenty of talented people cannot succeed for many reasons, and plenty of not-so-talented people find themselves successful beyond their wildest dreams.

The world is random, even cruel, and does not always reward merit or hard work or skill. Sometimes it does, but not always. Still, perhaps a more viable distinction to make is not between hard work and luck, but between what is up to us and what is not up to us. This is the distinction the Stoics tried to make back in the Third Century B.C. Pioneering novel ways to win—that's up to us. Being rewarded for hard work, such as winning a Super Bowl, is not. A season decides which teams compete for that chance. The media might even play a role in that. How players react to all of it is the real question.

Becoming an expert in a field—that's up to us. We do this by reading, by studying, and by going out and experiencing things. Getting hired as a coach for the New

England Patriots to teach that expertise is not. (Think of all the qualified people who weren't hired over the years because they were female or Jewish or African American.)

To write a prize-worthy work of literature—that's up to us. It's time in front of the keyboard. That's up to our genius. Being named as a finalist for the Pulitzer Prize is not. It's not that luck entirely decides these things, but it is very clear that other people do. I stated this in a previous chapter: You can be the most talented player in the world, have the smartest coach, and be surrounded by some of the best players yet still come up short of the goal—a national or world championship. You just need a little luck to get you over the top.

I want to give a shout-out now to so many teams that have never won a World Cup, world title, or national championship and let them know that help is on the way. Get up, Stretch your legs, click the website and don't forget to vote, my friends. theroadto59.com

"WE EACH HAVE OUR OWN BUCKET LIST, AND HOPEFULLY SOMEDAY WE CAN ALL EXPERIENCE THAT BUCKET LIST"

Chapter XI
Bucket List of Things to do in Sports

There are many events that loyal sports fans want to experience before they die—the so-called bucket list. These are the kinds of events that a father and son, or mother and daughter, can attend to remember a special time together. I have compiled a list of such events, which may be unattainable to attend for the average person but surely worthy of dreaming about before we pass.

Some events on my list might not make yours, but they are worth thinking about, if only as fantasies. The sporting event at the top of the list was difficult to come up with because I had to choose between the two biggest sporting event on the world stage. In the tradition of football games, I flipped a coin. The winner: the prestigious World Cup. Come on, man! Is there anything better than the World Cup? Well, maybe the Summer Olympics, the oldest sporting event in the world's history, but it lost the coin toss, so the number-two spot fell to the Olympics.

Between these two events, who does not want to party and witness the best athletes in the world compete? I wish I could have gone to Brazil, the party capital of the world, and be there at the same time as the 2014 World Cup. Of course, the Olympics is also a great place to watch elite athletes and beautiful people from around the world, and was also held in Brazil (Rio in 2016). And I also heard that Russia was a great place to be during the 2018 Winter Olympics. As you can see, I needed a coin flip to decide between the number-one and number-two spots on my list.

The number-three spot is easy: playing catch with your father or mother in the backyard. Those memories for a child and for the parent are priceless. As we travel down to the fourth spot, we gleefully play basketball with Michael Jordan. Well, okay, you'll need a time machine or else a rich imagination for that one—and for many of my top-twelve picks. Bugs Bunny had his time with Jordan; now it's time for you and me to do the same. I would also say that court-side seats at a Jordan game with your children would be acceptable.

So now we reach number five, and what do we have but another controversial pick at hand—the Masters. I know major PGA tournaments can be a blast because I once went to the Byron Nelson Classic in Dallas. To my surprise, it was one huge lawn party. I had a great time, and everyone else did, too. On TV, it looks so civilized, but that is far from the truth.

Okay, let's get to the number-six spot. Staying with the golf theme, this event stars you—sinking a hole-in-one, which would be pretty damn special. Right? When I reflect on it, I always think of the movie Tin Cup. Come on, man, he blows winning the tournament, sinks the hole-in-one after umpteen attempts, and still gets the girl. This is what I call a great day and a happy ending. I am just happy hearing a perfect crack when my club hits the ball.

Okay, my friends, let's travel now to the NFL and my favorite sport. This is personal to me because I would love not only for my home team to go to the Super Bowl but to win it, and to watch it at our home stadium when they do. What sports fan would not love to see that scenario unfold for their favorite team? My Vikings came so close in 2018 but came up short. So, being clear, number seven on the list is going to the Super Bowl and watching your team win in your home stadium. The 2021 Tampa Bay Buccaneers gave some lucky fans out there a great treat!

Number eight is easy: meeting your favorite athlete, having a conversation, and taking a selfie with him or her.

Number nine is another easy one: having your child's favorite athlete show up to his or her birthday party as a surprise. Our number-ten spot also involves your child going to a Major League Baseball game with him or her and catching a foul ball (or seeing your child catch it) as it's shown on television. For a parent, it is a very special feeling to see your child that happy.

Number eleven is beating your favorite athlete in his or her sport. Who would not like to take MJ to task? Beating Michael Jordan one-on-one, or even in a game of HORSE, would be a dream come true. What about scoring on your favorite goalkeeper or stopping a shot on goal from Pelé or Ronaldo? Now, as an older person, any of those very unlikely accomplishments would be pure perfection, and I would believe that I earned it. I think that a child, however, would know and understand that their hero let them win.

The last event on the list is special because it is something I got to experience, and it has stayed with me for a long time. I believe that no matter who you are or why you are there, it would bring a smile to you, and if your family members accompany you, that will just make it even more of a special occasion. I think this number-twelve on the list is a classic: play a baseball game on the 'Field of Dreams' in Dyersville, Iowa. If you can't play a full game, then play catch with a loved one—someone you care about—and afterward, sit down with them over a beer or a soda and just take in the surroundings and enjoy the silence. I guarantee it will stay with you for a lifetime.

There are others I would have liked to add to the list, but I limited it to twelve. Sports bring us together in a way that no other person, place, or thing can. Let's make sure that is how it will always be.

I want to leave you with some personal experiences I have had with some of the greatest sports figures in the country. My first experience was in 1991 with my ex-girlfriend, who was in charge of a 'Field of Dreams' celebrity softball game. That is where I got to meet some of the greatest athletes in sports. This next piece is from Mike Conklin of the Chicago Tribune:

"Among the players were Reggie Jackson, Bob Feller, Bob Gibson, Gene Oliver, Joe Pepitone, Curt Flood, Jimmy Piersall, Lou Brock, Jay Johnstone, and Jose Cardenal. They were making their first trip to the field, and so was Joe Hoerner, a 14-year major-leaguer who pitched in the 1967 and '68 World Series for the Cardinals. His participation was special. Hoerner was born in Dubuque, Ia., just 20 miles away, and he was a left-handed pitching phenom on a state high school championship team for that Iowa city. The weekend also had youth clinics, card shows, and a memorabilia auction in Dubuque. There was even a drama workshop. They earmarked a portion of the revenue generated during the weekend for Easter Seals, Points of Light Foundation, Baseball Assistance Team (BAT), Dyersville Little League, and the Dyersville Beckman High School band."

"On Sunday, the Old-Timers faced Hundley's fantasy campers, who paid $2,400 to take part in three days of playing and practicing alongside the former big-leaguers. On Monday, the weekend concluded with the Heroes of Baseball facing the Hollywood All-Stars, a team of TV actors—Richard Dean Anderson, Larry Drake, Jason Priestley, Adrian Zmed, Kelsey Grammer etc.—from such shows as Matlock, L.A. Law, MacGyver, Cheers and Beverly Hills, 90210. Tickets for the game were $50."

"An Upper Deck official estimated over 100 sets of press credentials were is- sued for the weekend. They installed a special TV satellite for network feeds. The second floor of the Dyersville American Trust & Savings Bank was turned into Command Central. Hundreds of volunteers—including one farmer who used his tractor to pull a batting cage from Dubuque to Dyersville—got involved."

My ex invited me to come down there and hang out. I took one of my best friends, Vadim, with me and we headed down there. I had just gotten out of the military and was preparing to go to college, so it was a pleasant surprise to go to this event before school started. When we got there, it was around 11:00 at night, and Molly was a little stressed. She asked us if we could help her out with some events wherever she might need us. Of course we were going to help her. Never turn down a smart and beautiful woman when she asks for your help. That night, we went to the Field of Dreams, just the three of us, and we played catch, drank some beers, and talked about what we were all up to and what our plans for the future were. We were young, free, and ready for anything. I think we got to bed around 2 a.m.

The next morning, Molly asked us if we could drive around the White Sox owners or the GM and his staff. I am not sure which one, but we did, and they promised us tickets in the owner's box anytime we were in town. We never followed up with them as life goes on, you get busy, you forget, and time goes by. It was fun talking with them, though. So, as the morning progressed, we were back at the volunteer headquarters and poor Molly seemed more stressed than the day before. It turned out that about twenty people showed up wanting free tickets to the game that Reggie Jackson had promised them the day earlier at some celebrity golf game. From what Molly gathered, Mr. Jackson had a few too many drinks at the golf course and was promising tickets to anyone he talked to.

I had a run-in with Mr. Jackson myself later that day. Now I think he was a dominant player, and everyone can have bad days, and this was one of Mr. Jackson's bad days. Before the game started, a 10-year-old boy ran up to Mr. Jackson and asked for his autograph. Mr. Jackson's reply was, "I am not fucking signing anything, okay." You should have seen the boy's expression. It relayed genuine sadness and disappointment.

After I heard this, I could not stand by and let this go without some kind of response. Now Mr. Jackson is a giant man and a professional athlete, so my military experience told me I was at a big disadvantage. What could I do to make him realize what he had just done was wrong? Well, I was younger then, very good-looking, very confident, and a military man, so I believed I could do anything, and the path I chose was what I believed was my best option. It was easy and very effective. I just flirted with his date—some young, beautiful woman. It was so easy; I just paid attention to her, talked with her. He was on the field; she was in the stands. Easy-peasy. I could see on his face he was getting angry. I went to find the boy he swore at and took him around to the other athletes and actors to get their autographs, and made the boy the center of attention as best I could. Kelsey Grammer was nice to me and the boy. I did not know who he was, but he was nice to us, so I gave him a three-star rating for being a nice guy. Ha-ha. For me, it was just another day hanging out with friends.

The next encounter was with my favorite athlete, Kirby Puckett. He was a friend of a friend. He would hang out with us. I saw him maybe six times altogether. He gave us T-shirts, autographs, etc. He just spent time with us, which is uncommon for a famous athlete. It was a very sad day when he passed. My prayers and thoughts go out to his family always. A person can do a hundred good, sometimes great things in their lifetime and only be remembered for one or two bad things. Sometimes, we have to get past that.

My next encounter is brief, but somehow very memorable. I was down in San Antonio at Fat Tuesday having a beer when Charles Barkley walked in and walked upstairs to the balcony with his entourage. My friend, who was the manager, asked me if I wanted to go up there and meet him. I was like, "Hell, yes." He brought me up there and introduced me, and I went to shake his hand, but instead, he grabbed my head like a basketball. His hands were huge. He let go, and he shook my hand and bought me a couple of rounds. We talked for about ten minutes. He apologized for grabbing my head; it was just a joke. I did not mind. I think he accepted me because my friend was the manager, but mostly because I was in the military at the time, and that was what we talked about (and the very cute server who was serving us) for those ten minutes. I present these examples of my interactions with three superstar athletes because it shows us their human side. Mr. Jackson had a bad day with many people around him, and Charles Barkley was just having a good time after a game in San Antonio. At any time, the situations could have been reversed. Like any of us on any given day, sometimes we are lucky, and sometimes we are not.

There are a couple more interactions I would like to share with you because they just stand out, and well, they were just very enjoyable for me to experience. I am sure there are many, many other people in the world who have similar stories to tell. For now, it is my turn to tell the story and for you to enjoy it.

It was one night in October, November—not sure. It seemed like over ten years ago when my friend Ron and I had VIP tickets to a Minnesota Timberwolves game. His aunt's company had a box at the Target Center, and he secured two VIP tickets to go to the game. I was okay with that. I was tired from working all day, but I found some energy to say "fuck yeah!" As we were waiting for the shuttle, we started a conversation with this young lady who had the same game passes we did, so I mentioned to her that we were also going to the game, showed her our VIP passes, and as any good person from Minnesota would do, we continued the conversation. Her father asked us how we were getting to the game. We said we were going to take the light-rail to the Target Center. He told us we would probably be late for the tipoff. I said, "Well, that's okay, we don't have a choice." He grinned and asked us if we would like a ride to the game since all of us were going to the same place. I looked at my friend Ron, and we nodded to one another and then looked at this stranger and said yes and thank you so much, we appreciate it. As we stepped into his SUV, we talked about basketball, and my friend Ron sometimes said things he should not have said.

We were talking about Kevin Garnett and the trade to Boston. I always liked Kevin Garnett, but my friend was not so nice when he said some things about K.G. My friend said the Timberwolves made a mistake in trading him and that the Timberwolves sucked—and sucked for a long time. The stranger told us there is something we should know before we continued with this conversation. He told us he was business partners with Glen Taylor, the owner of the Minnesota Timberwolves. Wow, that was impressive, and he also had an exquisite daughter sitting next to me. Start to a good night! He told us that K.G. was a diva. I suppose that most rich, famous people are divas in private and put on a show for the rest of us in the public eye. There are exceptions, though, like Kirby Puckett, Paul Walker, Princess Diana, and I am sure a couple more out there.

We all ended up at the game on time, and we said our goodbyes. They went their way, and we went ours. To tell you the truth, I cannot remember if the Timberwolves won that game. Ron and I enjoyed the game from the box, and the free beer and food, so all was well.

The next encounter I had is when I was in ROTC at the University of Minnesota. They tasked us with doing the color guard at the Vikings-Packers game. That was an awesome time. I met Brett Favre, John Randle, and many players from both teams. I enjoyed meeting John Randle the most. He is very intense guy, but like a giant teddy bear, really. I was so nervous carrying the flag out onto the field during the national anthem. It worried me that I might drop the flag, but I did not, and all's well that ended well.

My next experience with sporting events and athletes was more luck than anything else. This also happened while I was in college back in the Nineties. I had four tickets to a Twins' game, somewhere in right field. I wanted to take this very cute girl who I have been talking to for a couple of months, and she agreed to go, but only if I could find a friend for her friend. A double date, sort of. I had four tickets, so no problem, and the first person who came to mind was my friend John. He had not had a date in a long time, so he seemed the perfect choice. We made plans to meet at the game right next to the ticket office. I worked that day, and I was going to be late, so I called my friend John to come by and pick up the tickets and meet the girls until I could leave work. Until I realized I'd left the tickets at my apartment!

The girls were already on their way to the stadium and waiting on us. I told John I would call him right back. I called the Twins' ticket office and told them about my situation, and asked if they could replace the tickets or if they had any tickets left. As I was speaking with the ticket agent, I had a patient come up to me and ask about her eye exam, and I told her to please wait. The ticket agent asked if I was too busy to talk, and I just said I had a patient ask me a question, but it was not a problem. She assumed I was a doctor, and her whole tone with me changed. She was much more receptive to what I wanted. I did not want the girls to wait by themselves outside, so the ticket agent said that they would escort both of them to their seats behind home plate, and for me and my friend John to come to the ticket office to get the other tickets to meet our wives inside the stadium.

Wait—she said "wives," and for a millisecond, I thought, should I say something or not? Okay, I am going with this. I said thank you, and when John and I arrived at the stadium, the girls were so impressed. They were telling us how the staff escorted them to their seats and brought them food and drinks, and how nice they were to them. We did not tell the girls what really happened; we just went along with it all night. The whole thing was just plain lucky for all of us, although both girls ended up being a little too crazy for me.

My last experience was meeting my very lucky friend, which led to my writing this book and helping some team win the next world championship. I have been lucky in my own life and have had some incredible experiences so far. I hope everyone out there can have the same or similar experiences in their lives. It makes a tremendous difference.

I now want to add a last-minute piece of information about my very lucky friend. He called me the other day to tell me a couple of things that were happening with him. First, he came down with COVID-19 a couple of months ago, and he survived. You could say he was very lucky. He had all the symptoms but did not progress into the ICU and was never put on a ventilator. Thank God! Go back to the beginning of the book and you will remember the line about the sniper saying it's better to be lucky sometimes than to always be good. I truly believe that. Second, he is back in New Jersey, New York, and the Philadelphia area. You already know how he hates the area, so I am guessing another championship blackout is happening as we speak. He told me he has been in the area since the beginning of August. He also told me that he contracted COVID-19 in New Jersey. I suspect that as long as he is there, there will be no championships for those teams in those three states. I asked him what if some team in those three states won the contest; would he go, and would they win a championship? He laughed and said, "I am a professional good-luck charm, thanks to you, so I take nothing personally." He also reminded me that when he is not working on an assignment, there are no championships—only when he takes an assignment does the good luck work—and also that the Yankees, Eagles, Giants, and Devils are on the list of fifty-eight previous championships.

Chapter XII

The World Wide Contest

The regions eligible for international sports are the member countries of FIFA and can be found on our list in chapter X. All American sports teams will include the sports leagues of the MLB, MLS, NBA, NFL, NHL, and WNBA. Also included are NCAA Division I schools. This is your chance to help your favorite team win that national or world championship. The rules are straightforward. There is a link on the website where you can vote. Go to theroadto59.com for further details and rules for the Road to 59 Contest.

Remember, one vote per book. The concept of the contest is direct: Whatever country or team has the most votes is where my friend will go take a work assignment, and then like every other time he has done this, a championship will follow. To win this contest, your team will need a little luck also.

There are two phases of this contest, and because of COVID-19, the dates might change. The first phase of voting is for the American sports leagues, which will last until August 1, 2022, and phase two will be for the World Cup, with the voting lasting until November 11, 2022. There is also a chance to be able to win cash prizes. With the pandemic, and the world crisis going on right now, we want to help as many people as we can. I want to wish everyone a good night and good luck.

Bibliography

Schippers, Michaela. Van Lange, Paul M. "The Psychological Benefits of Superstitious Rituals in Sports: Study Among Top Sports persons," Journal of Applied Psychology, Vol. 36, No. 10 (October 2006).

Calin-Jageman, Robert. Caldwell, Tracy. "Replication of The Superstition and Performance Study by Damisch, Stoberock, and Mussweiler (2010), " Social Psychology, Vol. 45, No.3 (May 2014).

Wiseman, Robert. "The Luck Factor," Skeptical Inquirer, Vol. 27, No. 3 (May/June 2003). According to this article (published in the Journal of Applied Social Psychology in 2006)

Photographs

1. Sports Illustrated, March 3, 1980: US hockey team defeating the Russians in the Winter Olympics

www.ingramcontent.com/pod-product-compliance
Lightning Source LLC
Chambersburg PA
CBHW070452100426
42743CB00010B/1594